WHAT IS AMBER?

watch blobs of resin ooze from cracks in the bark and slowly flow and drip down the trunk. It looks very similar to wax dripping down the side of a candle, although not so fast. Not all tree resins can form amber, as most get broken down and decay. Only two types of tree living today produce stable resins that could, with time, fossilize into amber. They are the Kauri pine (*Agathis australis*) (fig 5) of New Zealand and species of the legume *Hymenaea* (fig 6) in east Africa and south and central America. The resin produced by the Kauri pine is called Kauri gum, although it is not a true gum.

HOW IS AMBER PRODUCED?

Several factors affect the production of amber from resin, a process known as amberization. Once the resin is exuded it hardens. Resin contains liquids such as oils, acids and alcohols, including the aromatic compounds that produce the distinctive resinous smell – two highly aromatic resins are frankincense and myrrh. Scientists call these liquids volatiles and they dissipate and evaporate from the resin. The resin then undergoes a process known as polymerization, whereby the organic molecules join to form much larger ones called polymers. Hardened resin is known as copal (figs 7, 8). Copal becomes incorporated into soil and sediments, where it remains long after the tree dies. It continues to polymerize and lose volatiles until the resultant

amber is completely polymerized, has no volatiles and is inert.

Many scientists thought that time was important in the fossilization of resin to produce amber, and the amberization process was estimated as taking from 2–10 million years. However, it now appears that many more factors are involved. Most amber in deposits around the world was not formed where it is found – the copal or amber has been eroded from the soil, transported by rivers and deposited elsewhere. For instance, amber from Borneo is 12 million years old and comes from sand and clay sediments that were deposited in a deep ocean. The fossilized resin from Borneo that comes from beds of

sandstone is completely inert and undoubtedly amber. However, resin that comes from beds of clay still contains volatile components, which means that it is still copal. So, the types of sediment in which the resin is deposited is much more important than time for amber formation. But what is not so clear is the effect of water and sediment chemistry on the resin.

7 Flow of copal from east Africa, produced by Hymenaea. *(Length 165mm)*

8 Stalactite of New Zealand copal (Kauri gum). (Length 180mm)

THE PROPERTIES OF AMBER

Amber is described as an amorphous, polymeric glass (see Glossary).

CHEMICAL PROPERTIES

The polymers are cyclic hydrocarbons called terpenes. One of these, turpentine, is a volatile liquid obtained from tree resin and is often used as a solvent. Amber generally consists of around 79% carbon, 10% hydrogen and 11% oxygen, with a trace of sulphur. Amber can be partially dissolved in organic solvents, but it is not affected by most alcohols.

PHYSICAL PROPERTIES

Amber has a hardness of 2–3 on the Mohs scale, which means it is not very

hard (about as hard as a fingernail) and can be easily scratched. It has a specific gravity (s.g. or relative density) of 1.04–1.10, which is only slightly heavier than water (s.g. of 1.00). Some pieces of amber contain air bubbles, which lower the s.g. and allow the pieces to float – amber is readily transported by rivers and tides. It has a melting point range of 200–380°C, but it tends to turn black and burn, rather than melt. It may fluoresce blue under ultraviolet light and will produce static electricity if rubbed. Amber is warm to the touch and, if broken, can produce a conchoidal fracture, which looks like the surface of a shell.

VISUAL CHARACTERISTICS

Amber is very popular for making jewellery and there is considerable interest in the insects and other objects, or inclusions, trapped in it, particularly large and rare ones. Because of this, amber is valuable and often faked. Often, fakes have tell-tale features that

enable them to be identifed visually without resorting to damaging testing.

There are some features found only in amber. A lot of modern amber jewellery contains circular, radial cracks called sun spangles (fig 9), which are made by heating and cooling the amber. Baltic amber commonly contains hairs from the flowers of oak trees and black cracks filled with tiny pyrite crystals. These rarely occur in other ambers.

Insect inclusions that are only a couple of millimetres long are likely to be genuine. Larger genuine insects in Baltic amber are often partly covered in a white coating, which is a product of decay. These features indicate that the specimen is genuine, although their absence does not indicate that it is fake, as many pieces of amber are perfectly clear and without any inclusions.

There is one type of fake that can only be identified visually. These are known as carved or drilled forgeries, whereby a piece of genuine amber is hollowed out and an inclusion is inserted. Any tests carried out on the surrounding amber would indicate that it is genuine. The inclusions used tend to be large (more than 10mm long). Knowledge of modern animal groups and those found in amber is useful in identifying fakes such as these, as only a few insects recorded from amber belong to living species, the rest are extinct. A faked inclusion will belong to a living species and usually only a specialist can identify it. Close examination under a

9 Pendant containing circular cracks known as sun spangles. These cracks are common in jewellery and are artificially produced by heating the amber. (Length excluding rings 60mm)

10 Pressed amber, also known as ambroid. This substance is formed by fusing small pieces of amber together. (Length 50mm)

microscope should reveal planes running through the amber where it has been cut and hollowed out. However, there are also natural cracks and flow planes in amber that look similar.

Pressed amber, also known as ambroid, is commonly found in Victorian jewellery and as the stems of tobacco pipes. It is formed by fusing small pieces of amber together under high temperature and pressure. It can be clear or cloudy, or a mixture of clear and cloudy swirls (fig 10), and it can be a variety of colours. Yellow and orange pressed amber is very difficult to distinguish from unadulterated amber. It cannot really be called fake amber because it is made out of genuine amber.

11 *Composite necklace containing fake amber beads. The large yellow/orange bead is celluloid, the oval red beads are phenolic resin, the oval cloudy yellow beads are caesin, the clear orange beads and small faceted beads are glass. (Length of largest red beads 27mm)*

THE PILTDOWN FLY

A carved fake discovered in the collections of The Natural History Museum, London, has recently become famous and dubbed 'The Piltdown Fly' (fig 12). The specimen consists of a large housefly in a piece of Baltic amber and was first mentioned by the German entomologist Hermann Loew in 1850. Much later, in 1966, the famous German entomologist Willi Hennig examined it and identified it as belonging to the living species *Fannia scalaris*. This was important because *F. scalaris* is a very advanced fly and suggested that Baltic amber does not accurately reflect the fauna at that time, because other living species should also be preserved in it.

Then, in 1993, the specimen was re-examined using a binocular microscope and anglepoise lamp. The lamp put out some warmth, which caused a crack to suddenly appear around the fly. This was initially worrying but also suggested that there was something amiss. Close examination from the side revealed two planes, one running flat through the amber and the other around the fly. Clearly, someone had cut a piece of amber in half, hollowed out one side, inserted a fly and glued it back together. The fly has bright red eyes and an abdomen broken in an unnatural way, which also implicated it. Fakes like this are rare, as copal and plastic are more commonly used for inserting inclusions.

12 *'The Piltdown Fly'. This specimen is a carved fake in which a modern housefly has been inserted into a piece of Baltic amber. You can see lines around the fly where the amber has been cut in half and one side hollowed out. (Length of fly 7mm)*

FAKE AMBER

Several substances are used to imitate amber, but there are several tests that can distinguish between them.

AMBER SUBSTITUTES

The following substances are commonly encountered as fake amber:

- copal;
- glass;
- phenolic resin;
- celluloid;
- casein;
- modern plastics.

Phenolic resin, celluloid and casein are often seen in old, Victorian, Edwardian or Georgian necklaces (fig 11), whereas other plastics are used for more modern necklaces. Only copal and modern plastics are used to embed fake inclusions.

COPAL

Copal is often sold as amber because it naturally contains insects and other inclusions. If it is exposed to light and air it degrades to produce a network of small polygonal cracks (crazing) on the surface (figs 13, 14). This process also happens to amber but takes much longer and the amber turns dark orange, whereas copal remains yellow. Copal with natural inclusions usually comes from east Africa, Colombia and the Dominican Republic. Most pieces are light yellow in colour and transparent. Some pieces from east Africa and Colombia can also be colourless or orange, and some pieces from Colombia and the Dominican Republic also have a brownish tinge.

The pieces can contain many different types of insect (fig 15). If some of the insects are only a couple of millimetres long, then the pieces are likely to be genuine. Again, knowledge of the differences between modern insects and those found in amber is useful as most insects in copal belong to living species. One good indicator is the proportion of flies with long or short antennae. Flies with long antennae are very common in amber whereas those with short antennae are rare. The opposite is true for the modern fauna – most flies trapped in copal have short antennae. There are several specimens in The Natural History Museum, London, that were labelled 'amber' but are actually east African copal. Many were originally shown as amber in the *Quarterly Journal of Science* in 1868. Another specimen labelled as Baltic amber was identified as belonging to the

13 East African copal with characteristic crazed surface caused by oxidation. The inclusion is a beetle. (Length of specimen 70mm)

14 New Zealand copal (Kauri gum) with crazed surface. (Length 145mm)

15 East African copal with insects. The three large flies belong to a family that has not been found in amber. (Length of specimen 53mm)

living cockroach species *Euthyrrapha pacifica* in 1911, but this again is in east African copal and not amber.

Faked inclusions are also found in copal. Copal has a much lower melting point than amber (less than 150°C) and melts rather than burns, allowing organisms to be inserted. Most copal fakes are large (more than 20mm in length) consisting of lizards, large crickets, spiders and butterflies. Often, the fakes contain only one large inclusion, which is in the centre of the piece with its legs neatly arranged (figs 16, 17). Kauri gum was commonly used as a medium for faked inclusions in the early 20th century. It is usually transparent and either yellow, orange or red with a crazed surface. Natural inclusions in Kauri gum are very rare as few insects live on the Kauri pine.

GLASS

Glass is easy to distinguish from amber because it is cold to the touch, dense and does not scratch with steel. See fig 11 for examples of fakes using glass and the next three substances.

PHENOLIC RESIN

Phenolic resin, which is also known as bakelite, is the most common material to be encountered in fake amber necklaces. The necklaces are usually made up of very large oval beads that are smooth or faceted. The beads get progressively larger the further they are away from the catch. There are two colours that are commonly encountered. The most common is a dark red with every bead virtually the same colour. The beads can be transparent or cloudy. Phenolic resin beads can also be cloudy yellow with swirls, which makes it harder to distinguish from genuine or pressed amber and requires physical or chemical analysis. Phenolic resin is slightly denser than amber.

CELLULOID

Celluloid (cellulose nitrate) is usually yellow and cloudy. It is difficult to distinguish from amber except that it is slightly denser and much more inflammable.

17 Iridescent beetle that has been inserted into melted copal. This species only lives in Asia whereas the copal probably came from east Africa. (Length of piece 19mm)

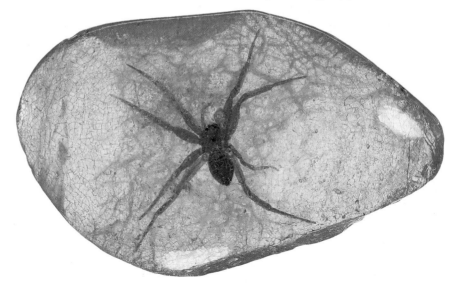

16 Spider that has been inserted into melted Kauri gum. Its large size and outstretched legs are tell-tale signs that it is a fake. Genuine spiders in amber usually have their legs tightly curled up under their bodies. (Length of piece 80mm)

FAKE AMBER

CASEIN

Casein is a plastic made from milk. The beads are cloudy and a dirty yellow colour, and it is also slightly heavier than amber.

MODERN PLASTICS

Modern plastics such as polyester and polystyrene are commonly used today to fake amber and to produce faked inclusions. These are harder to detect because some forgers can produce perfectly transparent pieces that are amber coloured. They are commonly used in Mexico and the Dominican Republic. As with copal, the faked inclusions are usually obvious, that is the inclusion is large (usually more than 10mm long), neatly centred in the piece and belongs to a living species (fig 18). Some forgers also add dirt to try to make it look more authentic.

In Mexico, green amber does occur but is extremely rare, yet necklaces and pieces made of green plastic are commonly sold as amber. In the Dominican Republic genuine lizards and scorpions in amber do occur but they are extremely rare and very valuable. It is very easy to visit there and buy a fake lizard or scorpion in plastic and pay a lot of money for it. The local lizard population is probably under threat from amber forgers. Plastic has also turned up in carvings that are supposed to be Chinese amber.

18 Lizard that has been inserted into plastic. Fakes like this are very common because genuine lizards in amber are extremely rare and valuable. (Length 75mm)

TESTS FOR AMBER

There are several tests that can easily be done to try to determine whether a specimen or necklace is genuine amber. These tests are potentially destructive and should only be used as a last resort if it is not obvious visually whether the piece is genuine or fake. For a summary of these see Table 1.

THE ALCOHOL TEST

An effective test to distinguish between amber and copal is the alcohol test. This involves putting a drop or two of alcohol, either isopropanol or ethanol, on a polished surface of the specimen and letting it evaporate. The volatile components that are present in copal react with the alcohol, which removes the polish and makes the surface sticky. If you press a finger against the spot a fingermark is often left on the surface of the copal (this can be polished away by rubbing it hard on a piece of cloth). There is no reaction with amber or the other fake materials and the polished surface is usually unaffected.

THE SCRATCH TEST

A good test to distinguish between amber and glass is to try to scratch it with a pin. If it does not scratch then it is glass (hardness of 5-6 on the Mohs scale). If it is amber or the other fake materials it will scratch, so choose part of the specimen where it is not noticeable and try to make the scratch as small as possible. Close examination of the specimen beforehand may indicate that it is not glass, without the need to use this test. If the surface is crazed then it is not glass. Glass can be scratched by diamond (hardness of 10 on the Mohs scale), so if the piece is already scratched this does not necessarily mean it is not glass.

THE HOT WIRE TEST

The hot wire test can be used to distinguish between amber and fake materials. This involves heating up a wire or needle red hot, allowing it to cool slightly and gently pressing the tip against the surface, which produces a puff of smoke. Sniffing the smoke can indicate what it is. Amber has a slightly acrid resinous smell and copal has a sweet resinous smell, whereas many artificial materials have an acrid plastic-like smell. The wire makes a mark on the surface, so choose part of the piece where it will not show. Take care when using this test because celluloid is highly inflammable and other plastics give off potentially harmful vapours.

THE SALTWATER TEST

Another test is to use a saturated salt solution to see if a specimen will float or sink.
• Fill a glass with half a pint (284 millilitres) of water.
• Add seven heaped teaspoons of table salt then stir. The water will initially turn cloudy, then gradually become clear as the salt dissolves.

• Leave it for several minutes, giving it a quick stir every 30 seconds.
• To test a necklace, remove one of the beads from the string and drop the bead into the glass. The salt solution has a higher specific gravity (1.1) than amber and copal so these materials will float, whereas glass, phenolic resins, celluloid, casein and some other plastics are denser than water and will sink. Polystyrene, however, has the same specific gravity as amber and will float.
• Always thoroughly wash the piece in water afterwards.

This test is of no use for amber pendants with a metal clasp as the weight of the clasp will make the pendant sink. Do not use this test if there are any cracks in the specimen as there is a danger that the salt water will penetrate the crack and, when it dries out, could produce salt crystals that enlarge the crack.

ANALYTICAL TESTS

There are several scientific tests used to identify amber and the two most often employed are infrared spectroscopy and mass spectrometry. Both produce a graph with many peaks reflecting the chemical composition of the sample. From studying the graph it is easy to find out whether the sample analysed is amber or one of the fake materials. Infrared spectroscopy can indicate whether the amber is Baltic amber or not, as Baltic amber has a characteristic plateau on the side of one of the peaks,

known as the 'Baltic shoulder' (fig 19). Mass spectrometry is able to distinguish between the different types of amber, but this process is more expensive and not widely available.

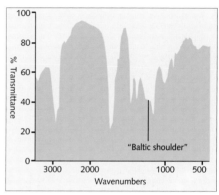

19 Infrared spectrograph of Baltic amber, showing the characteristic Baltic shoulder.

A Does alcohol make it sticky?
B Can it be scratched?
C Does it float in a saturated salt solution?
D Does a hot wire produce a resinous smell?

	A	B	C	D
Amber	N	Y	Y	Y
Copal	Y	Y	Y	Y
Glass	N	N	N	N
Phenolic resin	N	Y	N	N
Celluloid	N	Y	N	N
Casein	N	Y	N	N
Other plastics	N	Y	N	N
Polystyrene	N	Y	Y	N

(Y = yes; N = no)

Table 1. The results of four tests on amber and substances used to fake it.

THE USES OF AMBER

The most common use for amber is in jewellery. It is popular because it is light, warm to the touch and beautiful to look at. Most amber jewellery consists of pendants, necklaces, earings, brooches and rings (figs 20, 21, 22).

Because amber is fairly soft, it is easy to carve. Since prehistoric times it has been carved into a variety of ornaments, mainly of people or animals (figs 23, 24, 25), some of which had a significant meaning. It has also been used for everyday items such as bowls, cups, bottles, snuffboxes, knife handles, candlestick holders and quill pen holders. It has been used for recreational pursuits such as games boards, even with amber chessmen and dice (fig 26). Pieces have been skillfully inlaid in the surface of various items of furniture including chests, caskets and cabinets – even a chair has been made out of it. Amber has been used a lot for religious items. The main use in medieval times was for rosaries; later it was used to make altars and crucifixes. Amber has also been used for pipe stems and cigarette holders, although ambroid is more commonly used for these because it is slightly tougher (fig 27). More information on the many uses of amber can be found in the books listed on page 73.

THE AMBER ROOM

The most remarkable use of amber was when it was used to completely decorate a room. The famous amber room was commisioned by King Frederick I of Prussia in 1701. It was made from large panels inlaid with carefully shaped pieces of amber to form a mosaic. In 1717 it was given to Czar Peter the Great of Russia and placed in the Old Winter Palace in St Petersburg. In 1755

20 A beautiful Baltic amber pendant surrounded by diamonds, containing a spider and a cricket. (Length excluding ring 43mm)

21 A Baltic amber ring containing a long-legged fly. (Length of ring 19mm)

22 Necklace and earings made out of Baltic amber. The tiny beads are glass. (Length of the largest piece of amber in the necklace 41mm)

it was moved to the Ekaterininsky Palace in Tsarskoye Selo. During World War II, in 1941, the Germans invaded, dismantled the room and took it to Koenigsberg Castle. Since 1945 its exact whereabouts are unknown. There are a number of theories: it was either destroyed by bombing, was hidden somewhere, or is on a sunken boat at the bottom of the Baltic Sea. A single panel recently turned up in Bremen, Germany (in May 1997), but this has not cast any light on the whereabouts of the rest of the room. The Russians are now faithfully recreating the room based on old photographs.

AMBER PRODUCTS

Apart from decorative objects, amber has had other uses. The poorer grades of amber in the past were distilled to make colophony, amber acid (succinic acid) and amber oil. Colophony was dissolved in turpentine and linseed oil to make varnish. Amber varnish is hard, slow drying and dark, and was used for stringed instruments, horsedrawn carriages, ships' decks and in early photography. Copal and other natural tree resins were also used to make varnish in the past, whereas nearly all varnishes that are produced today are synthetic. Amber acid was used to produce soap, bath salts, pharmaceutical products, dyes and photographic chemicals, while amber oil was used in wood preservatives and insecticides. Both the acid and the oil were also used in the iron industry.

25 Crab carved out of Chinese amber. (Length 51mm)

26 Pair of Baltic amber dice. (Width 12.5mm)

24 Ancient artefact of Sicilian amber, showing a crude carving that would have been used as a talisman. (Length 65mm)

23 Plastic replica of a Stone Age horse carved out of a piece of Baltic amber from Poland. (Length 115mm)

27 Meerschaum pipe with pressed amber (ambroid) stem. (Length 103mm)

WHERE IS AMBER FOUND?

Amber occurs in many parts of the world, but most of the deposits are small and localized. Most of it is not of a high enough quality to make jewellery, but large deposits of jewellery quality amber are exploited commercially and most comes from the Baltic region and the Dominican Republic. All deposits are of interest to scientists who study their chemistry, while other scientists are interested in the deposits of amber that contain insects and other inclusions.

The main insect-bearing ambers come from the Baltic region, the Dominican Republic, Burma (Myanmar), Mexico, Lebanon, Siberia, Canada, New Jersey and Sicily. Insects have also been recorded in amber from Spain, Bitterfeld (Germany), China, Japan, Romania, Sakhalin Island (Russia), Claiborne (Arkansas, USA), Alaska and France. Recently, a few insects and other inclusions were also discovered in amber from the Isle of Wight, UK, and Borneo.

Fig 28 shows where these ambers occur and fig 29 illustrates their ages.

DATING AMBER

Amber can only be dated based on fossils in the associated sediments, but there is no way of knowing how long the amber took to become deposited. If the amber is re-worked (eroded from one deposit and re-deposited elsewhere) then the amber could be much older than the sediments suggest.

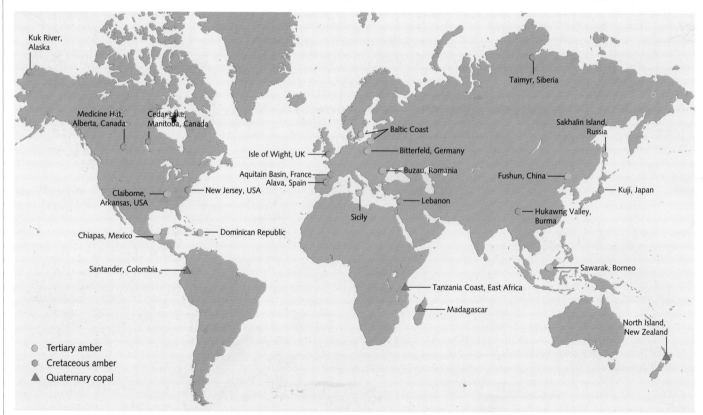

Kuk River, Alaska

Taimyr, Siberia

Medicine Hat, Alberta, Canada — Cedar Lake, Manitoba, Canada

Sakhalin Island, Russia

Baltic Coast

Bitterfeld, Germany

Isle of Wight, UK

Aquitain Basin, France — Alava, Spain

Buzau, Romania

Fushun, China

Kuji, Japan

Claiborne, Arkansas, USA — New Jersey, USA

Lebanon

Hukawng Valley, Burma

Sicily

Chiapas, Mexico — Dominican Republic

Santander, Colombia

Sawarak, Borneo

Tanzania Coast, East Africa

Madagascar

North Island, New Zealand

- ● Tertiary amber
- ● Cretaceous amber
- ▲ Quaternary copal

28 Map of the world showing the main amber and copal localities.

The oldest recorded amber comes from the Upper Carboniferous Coal Measures. Amber is also recorded from the Permian, Triassic and Jurassic periods. Plant and possible fungal inclusions have been reported from these early ambers but, as yet, no insects have been found. The oldest known insect-bearing ambers come from the Lower Cretaceous. The Cretaceous period was a very important time in the history of life on this planet – the flowering plants radiated and displaced the old conifers, cycads and ferns to become the dominant plants. At the end of the Cretaceous an extinction event occured that wiped out the dinosaurs.

SOURCES OF AMBER

Most amber is not of commercial interest so is unlikely to be encountered.

Of the commercial ambers, Baltic and Dominican are the best known and most readily available. These are dealt with in detail later. Some others do turn up from time to time and these are briefly described below.

ISLE OF WIGHT

Isle of Wight amber is of particular interest because it is from the Lower Cretaceous, which makes it one of the oldest insect-bearing ambers in the world (about 130 million years old). It is generally brown and cloudy with transparent yellow swirls, a lot of plant debris and some pyrite crystals (fig 30). A few similar pieces of amber have been found at Hastings, East Sussex, UK, but these have not revealed any insects (fig 31). This amber is slightly older (about 140 million years old).

30 Spider in Isle of Wight amber. (Length of longest leg 4.5mm)

31 Amber from Hastings, E. Sussex, UK. (Width 33mm)

32 Large piece of Burmese amber containing plant debris and nearly 200 insect inclusions. (Length of piece 100mm)

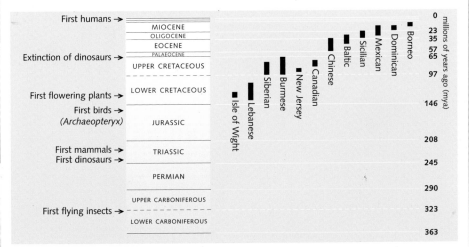

29 Geological time column showing some moments in evolution and the ages of some insect-bearing ambers. (Pliocene [5–2 mya] and Quaternary [2–0 mya] divisions not labelled.)

WHERE IS AMBER FOUND?

33 *The largest piece of Burmese amber in the world. (Length 500mm)*

34 *Piece of Lebanese amber. (Length 21mm)*

35 *Piece of Mexican amber. (Length 45mm)*

BURMA (MYANMAR)

Burmese amber (called burmite) is generally dark orange or red in colour. Pieces often contain plant debris (fig 32). Most of it was mined in the early part of this century from pits in the Hukawng Valley in northern Burma. Little has come from there since the 1930s. Burmese amber was thought to be Eocene in age, a date based on accompanying marine microfossils. However, a recent detailed examination of the collection at The Natural History Museum, London, showed that some of the insects belong to extinct families that are not known later than the Upper Cretaceous. It appears that the Burmese amber has been re-worked and is likely to be Upper Cretaceous in age. One piece of Burmese amber at The Natural History Museum is the largest piece of transparent amber in the world (fig 33).

MEXICO

Mexican amber is generally yellow with a brownish tinge, although red, green and blue forms also occur (fig 35). This amber often has parallel cracks running through it – a good indicator of genuine Mexican amber. It is mined in Chiapas State and is Upper Oligocene to Lower Miocene in age.

LEBANON

Lebanese amber is generally yellow. It contains a lot of cracks, which make it very fragile (fig 34). Similar amber has been found in Israel and Jordan. The insects in Lebanese amber are important because it is Lower Cretaceous in age.

36 *Piece of Sicilian amber containing two spiders. (Length 40mm)*

37 *Piece of very dark Sicilian amber. (Length 40mm)*

SICILY

Sicilian amber (called simetite) is usually orange or red, but can also be green, blue and black (figs 36, 37). It is usually transparent but can also be cloudy. When polished, the surface is highly reflective, almost dazzling. It is Oligocene in age.

CHINA

Chinese amber is usually transparent and orange or red in colour. Large pieces have been ornately carved (fig 38), but the source of these is unknown; they may have originally come from Burma (Myanmar). Amber with insects has been collected from Fushun in Liaoning Province and is Eocene in age.

BORNEO

Borneo amber is usually very dark red, almost black (fig 39), although yellow pieces also occur. The amber is generally slightly cloudy so any inclusions appear blurred. Some of the yellow pieces have not fully fossilized and are still copal. Borneo amber is Middle Miocene in age.

39 Piece of Borneo amber. (Length 53mm)

38 Carved piece of Chinese amber. (Length 120mm)

BALTIC AMBER

40 Map of the area around the Baltic Sea.

The most common amber in western Europe comes from the countries surrounding the Baltic Sea, namely Poland, Russia, Germany, Denmark and Lithuania (fig 40). It comes from a deposit known as the Blue Earth, which lies below the water table and extends out into the Baltic Sea. Storms can rip out amber from this bed, which dates from the Upper Eocene–Lower Oligocene, and wash it up onto the shores. Lumps of this amber are also occasionally washed up in Britain on the East Anglian, Kent and Yorkshire coasts after a storm (figs 41,42).

The largest Baltic amber deposits are on and around the Samland Peninsula, in an area that forms a small part of Russia between Poland and Lithuania.

Before World War II this area was known as East Prussia and formed part of Germany. The main city there is Kaliningrad, which was previously called Koenigsberg.

CHARACTERISTICS

When fresh, Baltic amber is usually lemon-yellow or orange and can be cloudy or clear (figs 43, 44, 48). The clear amber often contains insects, whereas the cloudy amber only rarely contains insects. However, because it is opaque, it isn't possible to see inside it. The cloudiness is caused by millions of microscopic air bubbles. A high proportion of bubbles can make a piece white and this is known as bone amber. With time – perhaps at least 50 years –

41 Large polished piece of Baltic amber dredged off the coast at Yarmouth, Norfolk, UK. (Length 170mm, weight 1048g)

42 Large piece of red Baltic amber washed up on Cromer beach, Norfolk, UK. (Length 205mm, weight 793g)

the clear amber oxidizes dark orange or red due to exposure to air or sea water (figs 45, 46). After a much longer time it develops a cracked crust such as is seen in some archaeological specimens. Most Baltic amber contains a lot of succinic acid (3–8%) and has been given the name succinite. This acid produces the characteristic 'Baltic shoulder' on the infrared spectrograph (fig 19). Other ambers that do not contain succinic acid are known as retinites.

There are a few ways of visually identifying clear Baltic amber. It commonly contains tiny hairs (1–2mm long) that probably came from the male flowers of oak trees (fig 47), although the resin itself came from conifers. The

44 Polished piece of cloudy Baltic amber. (Length 55mm)

46 Piece of oxidized orange Baltic amber. The curved, smooth ridges on the surface are where the piece has broken. This is known as a conchoidal fracture. Part of the natural unpolished crust can be seen, represented by small polygonal cracks. (Length 85mm)

43 Unpolished piece of Baltic amber from the Blue Earth, Kaliningrad, Russia. (Length 70mm)

45 Piece of highly oxidized red Baltic amber containing a spider and crane-fly. (Length 30mm)

47 Clusters of tiny hairs, probably from the male flowers of oak trees. (Length of the long hair in the centre 1.5mm)

BALTIC AMBER

presence of these hairs indicates that most of the clear resin was produced during the spring and summer, while the oaks were flowering. Often, Baltic amber also contains cracks, some of which are black and filled with lots of tiny pyrite crystals (fig 49). Some of the inclusions are also preserved in pyrite (fig 50). The oak hairs and cracks filled with pyrite crystals are rarely found in other ambers. The insect inclusions are often preserved with a white coating, which is formed from liquids that have escaped from the decaying insect's body and then entered the surrounding amber, turning it cloudy (fig 49). You can only find evidence for this process in Baltic amber.

49 Caddis-fly covered with a white coating formed as a product of decay. On either side of the insect is a crack filled with crystals of iron pyrites (pyrite). (Length 9mm)

50 Centipede in Baltic amber. This centipede has been half polished away and can be seen to be filled with pyrite crystals. (Length of piece 19mm)

THE BALTIC AMBER TREE

There has been a lengthy debate as to what type of tree produced Baltic amber. There is no doubt that it was produced by a conifer, but what kind? The extinct amber-producing tree was named *Pinites succinifer* in 1836 by H. R. Goeppert based on bark inclusions in the amber. It was later placed in the living genus *Pinus* (family Pinaceae). More recent chemical analysis indicated that Baltic amber is more similar to the resin produced by monkey-puzzle trees (family Araucariaceae, which includes the Kauri pine), particularly due to the presence of succinic acid. However, some trees within the pine family can also produce succinic acid, and araucarian inclusions are much rarer than pine inclusions. The Baltic amber producing tree had features of both pines and monkey puzzle trees. The plant and insect inclusions preserved in Baltic amber indicate that it was produced in a subtropical forest with a mixture of different trees.

48 Polished piece of fresh Baltic amber. (Length 70mm)

ARTEFACTS

Baltic amber has a long history. Amber artefacts have been recovered from many archaeological sites. The oldest British finds come from Gough's cave (around 10,000 BC) in Cheddar and Star Carr, Yorkshire (around 8,000 BC) (fig 51). This amber probably originated from the English coast. One of the most important archaeological finds is a Bronze Age amber cup that was discovered in a tumulus grave in Hove, Sussex, UK (fig 52). Other finds occur in countries that are a long way from the source of Baltic amber, indicating that trade in this substance started very early on. Certainly by 1,600 BC, a trade route had opened up to Greece.

In medieval times amber was collected by local peasants from the shore around the Samland Peninsula or netted from the sea itself. It wasn't until the 1850s that systematic mining began in the area with the formation of the Stantien & Becker company. They started by dredging the sea then went on to open-cast mining followed by underground mining on the Samland Peninsula. The company established a museum to house a large collection of amber, which was maintained by Professor Richard Klebs. In 1899 the company was purchased by the Prussian state and most of Klebs's collection passed to Koenigsberg University. Subsequently, 200 of Klebs's specimens were acquired by The Natural History Museum, London. Commercial open-cast mining on the Samland Peninsula continues to this day.

More information on the history of amber and amber mining can be found in the books listed on page 73.

51 Stone Age amber artefact from Yorkshire, UK. This piece is about 10,000 years old (8,000 BC). (Length 23mm)

52 Bronze Age amber cup from Hove, Sussex, UK. (Diameter 90mm)

DOMINICAN AMBER

Dominican amber comes from the Dominican Republic on the island of Hispaniola in the Greater Antilles (Caribbean, West Indies), not the island of Dominica in the Lesser Antilles. Although this amber was first mentioned by Christopher Columbus, interest in it only really started in 1960. There are many mines on the island, most of which occur in a mountain range called the Cordillera

53 Piece of Dominican amber containing a cockroach and two termites. The cockroach is on the left and the termites are close together on the right. (Length 53mm)

Septentrional. They consist of pits or tunnels that are mined by hand by the local inhabitants. Occasionally, the tunnels fill with water and collapse. The amber is taken to the town of Santiago, then to Santo Domingo, where it is checked by the amber museum before being sold by dealers. Unfortunately, by the time it reaches the open market the information on which mine it came from is lost. Suggestions as to its age have ranged from Upper Eocene to Middle Miocene, but current evidence indicates that most of the insect-bearing amber was deposited during the Lower to Middle Miocene. A couple of mines produce copal of much younger age (fig 55), although this does contain some extinct species.

CHARACTERISTICS

Dominican amber is commonly sold commercially both as jewellery and as pieces with inclusions. It is usually perfectly clear and comes in a variety of colours. Many pieces show a range of shades of yellow and orange, formed by a series of flows of slightly different colours (fig 53). It can also be green and blue (fig 54), although both these colours tend to fade with time. Dominican amber can contain cracks, air bubbles and water droplets. Only rarely is it cloudy, or contains pyritized cracks, or stellate (star-shaped) hairs.

The insects are more diverse and generally better preserved than those in Baltic amber. There are many tropical forms present, indicating that the amber was produced in a tropical forest. So far, the insect fauna has not been as well studied as in Baltic amber.

Dominican amber is chemically similar to east African copal and was produced by the same type of legume tree belonging to the genus *Hymenaea*. Plant fossils in the amber indicate the tree that produced Dominican amber differed from species living today and has been named *Hymenaea protera*.

54 Piece of blue Dominican amber. (Length 85mm)

55 Piece of Dominican copal. Although it looks very similar to amber it can be distinguished using the alcohol test. (Length 53mm)

It is the work of many scientists to describe and name the animals and plants in the world, both living and fossil. Each animal and plant is given a genus and species name in Latin. This is known as binomial nomenclature and was established by Carolus Linnaeus (also known as Karl von Linné) in 1758.

A species is generally defined as a group of organisms that can only reproduce with each other. These organisms are alike and have a similar morphology, although you can get differences between the sexes, which is known as sexual dimorphism.

TAXONOMY

The process of describing and naming genera and species is known as taxonomy. A specimen that is used for the basis of a species description is known as the type. The different species are placed in a classification that consists of a hierarchy of taxonomic units, known as taxa (see Panel). However, only a few of the taxa are used regularly. The most common units are: Phylum, Class, Order, Family, Genus and Species.

EVOLUTION AND EXTINCTION

A population is a large number of individuals of the same species living in a given area. The populations of a species define its biogeographical distribution. Populations can often split up or merge due to changes in the environment. If a population splits and the subsequent populations are then separated for some time, subtle changes can take place over the generations, and evolution takes place. If the populations are separated for long enough, they may change so significantly that if the populations merge they can no longer reproduce with each other. This means a new species has evolved, a process known as speciation.

Environmental changes put pressures on a species. It copes with these either by adapting and evolving or migrating to a more favourable area. If the environmental changes are rapid then it may not be able to cope, in which case the species will die out and become extinct. Some species adapt to changes better than others and some groups evolve faster than others. Species that have strict ecological requirements are more prone to extinction than those that do not. Species that are still living today are described as extant.

Though many species have been described and named, there are still many more that are waiting to be discovered. Insects have a high diversity and about a million extant species of insect have been described and named, but it is estimated that more than 10 million species are living on this planet today. Only about 20,000 fossil (extinct) insect species have been described and named but many more (estimates suggest 100,000) await to be described.

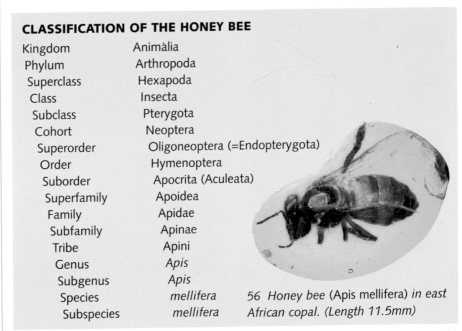

CLASSIFICATION OF THE HONEY BEE

Kingdom	Animàlia
Phylum	Arthropoda
Superclass	Hexapoda
Class	Insecta
Subclass	Pterygota
Cohort	Neoptera
Superorder	Oligoneoptera (=Endopterygota)
Order	Hymenoptera
Suborder	Apocrita (Aculeata)
Superfamily	Apoidea
Family	Apidae
Subfamily	Apinae
Tribe	Apini
Genus	*Apis*
Subgenus	*Apis*
Species	*mellifera*
Subspecies	*mellifera*

56 Honey bee (Apis mellifera) *in east African copal. (Length 11.5mm)*

AMBER INCLUSIONS

Inclusions are all the objects that are trapped in amber. The most familiar of these are insects and spiders, but there are many other things that have also been trapped. These include bacteria, fungi and many different types of plants. Invertebrate animals (those without backbones), other than insects and spiders, include worms (fig 58), snails (fig 59), rotifers, tardigrades and microscopic protozoa. You can also find vertebrates, but these are extremely rare and consist of frogs, lizards (fig 57), birds' feathers and mammal remains. The inclusions are very important for studying past diversity, ecology and biogeographical distributions.

There is a bias in size with the animals that are trapped in amber. Animals that are larger than 20mm long, such as lizards, frogs, scorpions, large insects and spiders, are generally strong enough to pull themselves free of the sticky resin, so the majority of inclusions that are trapped are much smaller – only a few millimetres long. For example, the largest natural inclusions in amber in the collections at The Natural History Museum, in London, are about 20mm long (fig 60).

57 Lizard in Dominican amber. (Length 55mm)

58 Roundworm (nematode) in Baltic amber. (Length 2.6mm)

59 Snail in Burmese amber. The small black inclusions are its droppings. (Length of piece 12mm)

60 Alder fly in Baltic amber. (Length 21mm)

61 Planthopper bug in Baltic amber, with concentric lines around it, showing where it struggled. (Length 7mm)

EVIDENCE OF ANIMAL BEHAVIOUR

Amber can also preserve evidence of the behaviour of animals before and after they are trapped. When an animal gets trapped in resin it is still alive, so there is sometimes evidence of a struggle, such as concentric lines in the amber around them (fig 61). Some flies and harvestman spiders are able to break off their legs to enable their escape. Isolated legs are often seen in amber, as well as flies with some of their legs lying nearby (fig 62). Some flies are trapped while still mating (fig 63), whereas some insects lay eggs in amber just before dying. Other insects are incomplete and their struggle on the surface of the resin may have brought them to the attention of larger animals looking for a meal. A few inclusions in amber are mouldy and the mould probably formed after the animal died on the surface of the resin, before it became completely engulfed (fig 64). Of particular interest is where there is direct evidence of two types of animal interacting. This will be discussed in more detail later (see page 26).

ANIMAL PRODUCTS

The products of insects and other organisms are also trapped in amber. Insect droppings are very common in amber (particularly Dominican) and are small, black and barrel-shaped (fig 65). Sawdust plugs also occur, which would have been pushed out of holes by wood-boring beetles. Strands of silk

62 Crane-fly in Baltic amber, which has broken off its legs in an effort to escape. (Length 3.5mm)

64 Mouldy scuttle fly in Dominican amber. (Length including mould 3.6mm)

63 Caught in the act – a pair of mating scavenger flies in Dominican amber. (Length of large fly 1.8mm)

23

AMBER INCLUSIONS

65 *Insect droppings in Dominican amber.*
(Length of topmost dropping 1.9mm)

from a spider's web are seen occasionally and even spider's nests are known. When an insect or spider grows, it has to moult and shed its skin: often the skins have been preserved (figs 66, 67). Pieces of shed lizard or snake skin have also been found (fig 68). Ants and termites throw a lot of debris out of their nests, and this can become trapped in amber, although this is not easily distinguishable from general debris. Ants eat other insects so their waste also includes lots of bits of insects. A vertebrate dropping or regurgitated pellet was recently discovered in Baltic amber (fig 69). It consists of a concentration of beetle bits glued together. No-one is sure what produced it – whether it came from a bird, lizard or even a bat.

PRESERVATION

The chemical properties of the different types of amber preserve the organisms in different ways. Dominican amber is the best preservative, with most insects appearing in perfect condition, including their internal tissues. Many Baltic amber insects are not as well preserved and often have a white coating around them. Most are also hollow due to decay, with little or no tissue preserved, whilst others are filled with pyrite crystals, which can also penetrate the wing membranes turning them black. The insects in Burmese, Mexican and Borneo amber are generally less well preserved. Often they are semi-transparent, probably due to complete penetration by the resin, and may be partially dissolved, as many are incomplete and distorted (fig 70).

67 *A shed earwig skin in Baltic amber.*
This specimen has very long pincers.
(Length 9mm)

66 *A shed caterpillar skin in Dominican amber. (Length 3.4mm)*

INORGANIC INCLUSIONS

It is not only animals and plants that get trapped in amber. Bubbles and water droplets are also common. Occasionally, you can see water droplets that also have a bubble of air inside them. When the amber is turned, the bubble moves so that it stays at the top, acting as a natural spirit level (fig 71). There is a debate as to whether amber is a perfect sealant. This is important when considering whether the bubbles in amber are actually preserving ancient atmospheres. Since the proportions of gases in the atmosphere have probably changed through time, some scientists believe that amber bubbles do reflect ancient atmospheres, whereas other scientists consider that small molecules can migrate through amber. This would alter the proportions of gases in the bubbles. It is likely that oxygen in the bubbles would have reacted with the amber, thus altering the composition of the air in the bubbles and so they would not accurately represent the atmosphere from long ago. Apart from the inclusions, flow planes and cracks also occur in amber. The flow planes indicate successive periods of resin flowing over each other. Sometimes they are crazed due to the surface of the resin hardening before the next flow covered it (fig 72). The cracks would have developed after the amber formed, probably due to pressure generated by the weight of overlying sediments (or by the mining and polishing processes).

70 Distorted planthopper bug in Mexican amber. (Length excluding wings 2.7mm)

71 Air bubbles inside water drops in Baltic amber. (Length of longest drop 0.9mm)

68 Piece of shed lizard or snake skin in Burmese amber with a mite lying over it. (Length of mite 1.5mm)

69 Vertebrate dropping or regurgitated pellet in Baltic amber, consisting of beetle remains. It could have been produced by a bird, lizard or bat. (Length 6mm)

72 Crazed flow plane where the surface dried out prior to the next flow.

ANIMAL INTERACTIONS

Some amber specimens show evidence of behaviour that cannot otherwise be seen in the fossil record. Of particular interest is where a piece of amber contains direct evidence of two or more animals interacting with each other, either still attached or in close proximity. There are four types: parasitism, mutualism, commensalism and predation.

PARASITISM

Parasitism is where an organism benefits but the other (the host) suffers. There are many insects in amber that would have been parasitic. This can be inferred from the fact that their living relatives are parasitic today, such as many wasps. The winged adult lays its eggs in the host, which can be an adult insect (or spider), pupa, larva or egg. The wasp larva (the parasite) hatches out and slowly devours its living host. Although there are many different species of wasp in amber, usually there is no way of knowing which other organisms they parasitized. The only way of knowing for sure is when you get a specimen with the parasite still attached to its host. There are rare examples of bugs in Dominican amber that have a sac attached to them with a developing wasp larva inside (fig 73). Occasionally you can get an indication of what animal an organism parasitized when you find several associated insects in one piece of amber. There are clear examples in amber of parasitic mites still attached to flies, caddisflies and moths (fig 74). The earliest record of parasitism is in Lebanese amber. Some nematode worms are parasites inside insects. There are specimens of flies and other insects that have these worms coming out of their abdomens or lying in the amber nearby (fig 58).

MUTUALISM

Mutualism is a partnership where each organism benefits from the other. Although there are no direct examples of this in amber, there are specimens that imply mutualism. Termites have bacteria in their guts to help them digest cellulose. The bacteria produce methane as a byproduct of this process. Many termites in amber have large bubbles of methane projecting from their abdomens, which would have been produced by the bacteria after the termites were trapped. It's also possible that some methane was produced by the decay of the termite's internal tissues.

COMMENSALISM

Commensalism is where one organism benefits but the other is unaffected. One type of this is phoresy, where an animal hitches a lift on another. There are several examples of this in amber,

73 *Leafhopper bug in Dominican amber with a dryinid wasp sac attached to its head. (Length of bug excluding legs 2.9mm)*

74 *Fruit fly in Dominican amber with a parasitic mite attached to its abdomen. (Length of fly 2.9mm)*

75 *Snipe fly in Baltic amber with a pseudoscorpion hanging onto its legs. (Length of fly 8mm)*

particularly involving flightless arachnids getting rides from flying insects. Pseudoscorpions can be seen using their pincers to hang onto the legs and bodies of flies and wasps in Baltic amber (fig 75), and onto beetles in Dominican amber. Phoretic mites can be seen on flies, bees, beetles and termites in Dominican amber, although it is often difficult to tell whether mites that are associated with an insect are phoretic or parasitic. Juvenile nematode worms have been recorded on ants and beetles; and a beetle larva has been recorded attached to the head of a bee, both in Dominican amber.

PREDATION

Predation is where one animal eats another. Much of the evidence is inferred because the type of insect observed in amber is wholy predatory today. Some insects in amber have been observed with other insects in their jaws. This is either due to one catching the other to eat just before entrapment, or it is using its jaws to grab hold of something to try to pull itself free of the sticky resin (fig 76). Flightless female scuttle flies have been found in Dominican amber. Today, they are mainly scavengers in termites' and ants' nests, but some are specialized predators. The female crawls down into a nest, lays its eggs and crawls out. To evade detection she secretes the same chemicals that the ants and termites use to recognise each other. If it is a

predatory species, then when the larvae hatch out they devour the termites, ants or their larvae. One scuttle fly occurs in a piece of Dominican amber in association with two species of termite and a rove beetle, which also lives in termite nests (figs 77, 78, 79). It is therefore very likely that this species of scuttle fly lived in the nests of termites rather than those of ants.

76 Ants in Dominican amber. One is holding the other's abdomen with its jaws to try to pull itself free of the resin. (Length of ant on left 3.2mm)

77 Piece of Dominican amber containing termites, a flightless scuttle fly and a rove beetle. The fly and beetle would have lived inside temite nests. (Length of piece 20mm)

78 Close-up of the flightless scuttle fly in fig 77. (Length 1mm)

79 Close-up of the rove beetle in fig 77. (Length 1.3mm)

PLANT INCLUSIONS

The remains of plants are common in amber, although identifiable specimens of plant structures, such as leaves, twigs, cones and flowers are rare. The most common remains are fragments of bark and hairs from oak flowers (in Baltic amber). The bark fragments probably came from the trees that produced the amber. Pollen and spores also occur but they can only be seen by using a high-powered microscope. The Baltic amber flora has been well studied over the years, but little has been described from Dominican amber. Plant fossils in amber include mosses, liverworts, lichens, ferns, gymnosperms and angiosperms

(the flowering plants). In Baltic amber, most of the identifiable plant remains are gymnosperms or angiosperms, whereas the majority of specimens in Dominican amber are angiosperms.

GYMNOSPERMS

The gymnosperm specimens in Baltic amber are either from conifers or cycads. The conifers are the most common and consist of the twigs and cones of cypresses (fig 80), pines (fig 81), redwoods and podocarps. It is likely that some of the pine specimens came from the tree that produced Baltic amber, although the species (*Pinus*

81 Pine cone in Baltic amber. (Length 18mm)

succinifera) was described based on the microscopic examination of bark.

ANGIOSPERMS

The angiosperm (flowering plant) remains consist mainly of leaves (fig 82) and flowers (fig 85), and over sixty families have been recorded. Most angiosperm specimens are difficult to identify by a non-expert; however in Baltic amber there is one kind that is fairly common and easy to identify. As stated earlier, Baltic amber commonly contains clusters of tiny stellate (star-like) hairs. These are attributed to coming from the male flowers of oaks. Not only are the hairs preserved but so are isolated bracts (fig 84) and, rarely, complete flowers (fig 86) with the hairs still attached. The high abundance of oak hairs in Baltic amber indicates that most of the transparent resin was secreted during the spring and summer, while the oaks were flowering.

Other trees that would have grown in the Baltic amber forest (fig 87) include maples, holly, beeches, chestnuts,

80 Cypress twig (Thuja) in Baltic amber. (Length 13mm)

82 Angiosperm leaf (Eudaphniphyllum) in Baltic amber. (Length 28mm)

laurels, magnolias, proteas and willows. Trees represented only by pollen include birch, horse-chestnut and lime. Mistletoe grew on the trees.

Herbaceous and shrubby plants in Baltic amber include members of the palm, heather, *Geranium* and saxifrage families. In addition, there would have been some of the earliest grasses in the world, represented by two species.

Other familiar flowering plants in the Baltic amber forest, which are only represented by one species, belong to the arum, lily, flax, olive, elm, rose, rock rose, tea, carrot, and nettle families.

In Dominican amber, leaves (fig 83) and flowers have been found belonging to the legume *Hymenaea*, which is the tree that would have produced the amber. The Dominican amber tree, which differs from living species and is therefore extinct, was named *Hymenaea protera*.

84 Bract, probably from the male flower of an oak tree, in Baltic amber. (Length 9mm)

86 Male oak tree flower (Quercus) in Baltic amber. (Length 3.7mm)

85 Angiosperm flower (Croton) in Dominican amber. (Length 6mm)

83 Leaf of Hymenaea protera, *the tree that produced Dominican amber. (Length 52mm)*

THE ANCIENT AMBER FORESTS

The insects and other inclusions in amber can tell us a lot about the ecology of the ancient forests, a study known as palaeoecology.

The palaeoecology of the Baltic and Dominican amber-producing forests is better known than for any others. From the evidence available one can imagine what the forests were like (fig 87). The subtropical Baltic forest would have been a mixture of coniferous and deciduous trees. The pine trees would have secreted copious amounts of resin as flows running down the trunks, stalactites dripping from branches and hardened lumps stuck to the bark. Oaks would have been flowering with hairs and pollen from the flowers drifting in the warm breeze. You can imagine constant activity from the insects and other animals in the forest: caterpillars, crickets and stick insects munching on the leaves. Male crickets singing to attract mates. Columns of aphids (greenfly) on the twigs, each with its long proboscis inserted through the bark to suck the sap, constantly attended by ants that feed on their honeydew secretions and protect them from lacewings. You can imagine lines of ants marching up and down the trees in the forest.

The forest was likely to have been fairly dense, with clearings where old trees had died and rotted while other plants grew around them, competing for the sunlight. The many scavengers such as woodlice, earwigs, beetles, termites, ants, cockroaches, mites, barklice, millipedes, springtails, silverfish and bristletails would have crawled on the rotting logs, under the bark or in the leaf litter on the forest floor, munching on the decaying matter. Beetle grubs bored into the rotting wood to make a network of galleries, while female ichneumon wasps walked jerkily on the surface of the tree trunks, with their antennae waving, listening out for movement so that they could insert their ovipositors into the wood to lay their eggs in the grubs.

Mushrooms and other fungi growing on the logs and forest floor would have provided food for the larvae of fungus gnats. The gnats and midges would have formed mating swarms, and hoverflies would have flitted in and out of the sunbeams. Bees and wasps would have been buzzing around the flowers to pollinate them. Spiders would lie in wait in webs for their next struggling meal. Meanwhile, other predators such as centipedes, damselflies, harvestmen, praying mantises and various large flies actively hunted for their next meal or lay in wait for it to come along.

87 Artist's impression of the Baltic amber-producing forest.

The vertebrates, perhaps mammals, birds, lizards and frogs, would have fed on the various insects and plants, and in turn provided food for parasites including mosquitoes, black-flies, biting midges, horseflies, sandflies, fleas and lice. There would have been occasional ponds, fed by streams in the forest that supported the predatory nymphs of damselflies, which may have fed on wriggling midge larvae. Also in the ponds and streams were scavenging mayfly and stonefly nymphs, and caddis-fly larvae that protected themselves in portable houses of plant fragments. On particular days, mayflies would have hatched, swarmed above the ponds, laid their eggs and died.

The animal and plant species in the Dominican amber forest were different, but the groups would have lived in a similar way to those in the Baltic forest. The essential difference was that the Dominican forest was tropical and therefore hotter, more humid and supported a greater diversity of species. The majority of plants would have been angiosperms, some of which would have been epiphytes that lived in the canopy. Some insect groups – the ants and termites – were probably more abundant in the Dominican forest than in the Baltic forest, whereas others, such as greenfly, were much rarer.

THE SEARCH FOR DNA

DNA (deoxyribonucleic acid) is the molecule that contains the information necessary for the growth and function of all living organisms. It is made out of subunits called nucleotides, which are joined together to form a double helix (fig 88). Biologists in recent years have been extracting it from organisms to investigate the relationships between them. In the 1980s it was realized that cell structures are preserved in insects in amber, so then the search was on to find DNA in the insects' cells.

REPORTS OF DNA IN AMBER

DNA was first reported to have been recovered from amber in 1992 when scientists in California claimed to have extracted fragments of DNA from an extinct species of bee (*Proplebeia dominicana*) (fig 89) in Dominican amber. Shortly afterwards reports appeared of DNA from an extinct species of termite (*Mastotermes electrodominicus*) (fig 96), also in Dominican amber, by scientists in New York. This was followed by reports of DNA

88
The double helix structure of a DNA molecule.
32

extraction from a beetle in Lebanese amber. However, only small bits of the DNA string were recovered. There has been some scepticism as to whether these claims are genuine or the result of contamination. Experiments on the survival rate of DNA have shown that it breaks down very quickly, particularly in the presence of water. However, the insects in amber are dehydrated and if this happened quickly then it could possibly halt the decay of the DNA.

At The Natural History Museum in London, scientists have tried to repeat the experiments to obtain DNA from the Dominican amber bee *Proplebeia dominicana* (fig 89). These bees are common in Dominican amber because they collect resin to make their nests. Several suitable specimens were selected, broken up and tested, but no insect DNA was recovered. This casts doubt on the earlier reports because it appears that the experiments are not replicable, which is a fundamental requirement for reliable scientific results.

IS *JURASSIC PARK* POSSIBLE?

Even if DNA could be extracted from insects in amber, a real-life Jurassic Park is not possible. There are many reasons why such a venture will remain fiction. First, there are no known insect-bearing Jurassic ambers. Second, contrary to popular belief, mosquitoes (Diptera: Culicidae) (fig 90) are extremely rare in amber. There is one

recorded from Canadian amber, but this requires confirmation. There are only a handful known in Baltic amber and a few tens of specimens in Dominican amber. There are, however, other biting insects, known from Mesozoic deposits, that may have fed on dinosaurs. Black-flies (Diptera: Simuliidae) (fig 94) are known from Middle Jurassic deposits and two have been found in Cretaceous amber. The oldest horsefly (Diptera: Tabanidae) (fig 93) was found recently preserved in limestone of Lower Cretaceous age in Dorset, England, but none are known in Cretaceous amber. Biting midges (Diptera: Ceratopogonidae) (fig 92) have been found in Canadian, Siberian and Lebanese amber, but they can feed on the blood of many things, including other insects. A few are known in Canadian amber that have jaws adapted for biting vertebrates, however it is debatable whether they could have fed

89 A bee in Dominican amber, of the species Proplebeia dominicana, *which was used to attempt DNA extraction.*
(Length 4mm)

on dinosaurs. Sandflies (Diptera: Psychodidae) have been found in Burmese and Lebanese amber and one has large mouthparts similar to those of a living species that feeds on the blood of crocodiles (fig 91). This type of fly could well have fed on the blood of dinosaurs. However, it is extremely unlikely that anyone will destroy this specimen on the remote chance of extracting DNA because it is a male and only female sandflies feed on blood.

Even after an insect is trapped in resin, bacteria and enzymes continue working in the gut, rotting the insect from the inside. Indeed, many insects preserved in amber, particularly Baltic, are completely hollow without any

internal tissue preserved. If it is so difficult to get DNA from an insect in amber, then the chances of getting any DNA from something it fed on are even more remote. If it were possible to extract DNA from a blood meal in an amber insect, only tiny amounts of the entire DNA string (genome) would be recovered and it would probably be contaminated with bacterial and insect DNA. Key parts of the genome would be required to work out which type of animal the blood came from. Biologists could only guess at what was missing from the complete DNA string, without knowing for sure. Although scientists can manipulate and make copies of DNA, they can't make it grow into an animal.

91 Sandfly in Lebanese amber. This type of fly may have fed on the blood of dinosaurs! (Length 1.5mm)

92 Mating pair of biting midges in Baltic amber. (Length of male on right 1.6mm)

93 Horsefly in a piece of Baltic amber. (Length 15mm)

90 Mosquito in Dominican amber. (Length excluding upward-pointing proboscis 3.7mm)

BIOGEOGRAPHY

The study of the geographical distribution of animals and plants is called biogeography. Many entomologists study present-day insect species and their biogeographical distributions to work out how they evolved from other closely related species and also where this speciation probably took place. Some scientists have used the theory of plate tectonics to explain the occurrence of closely related species in different parts of the world. However, the study of insects preserved in amber is yielding important information on past changes in biogeographical distributions that cannot be deduced solely from looking at the distributions of present-day species. Indeed, some occurrences in amber are completely unexpected because their living relatives are found on the other side of the world today. Clearly, plate tectonics has had little or no effect on these distributions. They were probably controlled by environmental and/or ecological pressures, which made the species migrate from one area to another or made large populations shrink.

FLY DISTRIBUTIONS

There are many species of fly in Baltic amber whose nearest relatives live in south-east Asia and/or southern Africa today – they no longer live in Europe. For instance, the black-fly in figure 94 may belong to the subgenus *Morops*, which is only found today in south-east Asia. The snipe fly family Rachiceridae (fig 75) lives mainly in south-east Asia and North and South America today, with only one species living in central Spain. The closest relatives of the Baltic amber horsefly in figure 93 live today in Australia, south-east Africa, Chile and eastern USA. Similar surprising occurrences are being recognised in Dominican amber, such as scuttle flies belonging to the genus *Abaristophora* (fig 95). This genus only occurs today in Nepal and New Zealand.

TERMITE DISTRIBUTIONS

There is a primitive genus of termites called *Mastotermes* that lives only in Australia. This genus has turned up in Mexican and Dominican amber (fig 96), and many Eocene, Oligocene and Miocene fossil insect localities in Europe (including the Isle of Wight and Hampshire, England). So this termite must have had a worldwide distribution in the past, which has been reduced to a relict distribution in Australia (it has been accidentally introduced to New Guinea and New Zealand by man).

96 Large termite in Dominican amber. This termite belongs to the genus Mastotermes, *which is found today in Australia. (Length 26mm)*

94 Black-fly in Baltic amber. This fly may belong to the subgenus Morops, *which is only found today in south-east Asia. (Length 2.1mm)*

95 Scuttle flies (genus Abaristophora) *in Dominican amber. This genus is only found today in Nepal and New Zealand. (Length of fly on right 1.4mm)*

ARTHROPODS

The Phylum Arthropoda consists of all the invertebrates with jointed limbs and an exoskeleton generally made of chitin. The body is segmented and is usually divided into a head, thorax and abdomen, although in some groups the head and thorax are fused. They grow by moulting and often have compound eyes consisting of many lenses. This phylum contains five superclasses. Four are found in amber, the fifth is the extinct Trilobitomorpha (trilobites).

1. CRUSTACEA – CRABS AND THEIR RELATIVES

The Crustacea have a head-shield (carapace) and two pairs of antennae. There are many different groups of predominantly sea-dwelling Crustacea, including crabs, lobsters, shrimps, prawns, crayfish and barnacles. Only two orders belonging to the class Malacostraca have been

recorded in amber: the Amphipoda and Isopoda. Amphipods are laterally compressed (= flat sides) and are curved. They usually live in the sea or in freshwater and some are known commonly as sandhoppers. They are extremely rare and have been recorded in Baltic and Dominican amber. Isopods, by contrast, are dorso-ventrally flattened (= flat top) (fig 98). They live in the sea, freshwater or on land and are known commonly as woodlice or sea slaters. They are more common than the amphipods in amber but they are still very rare.

2. CHELICERATA – SPIDERS AND THEIR RELATIVES

The Chelicerata have their head and thorax fused together to form a cephalothorax, which has six pairs of appendages and no antennae. The first pair of appendages at the front of the head often form curved, downwards pointing fangs known as chelicerae. The second pair are known as pedipalps, which are variable in appearance and have either a sensory or food-gathering function. The other four pairs of appendages form the characteristic eight legs. This group contains two classes, the Arachnida (spiders, mites and scorpions) and the Merostomata (horseshoe crabs and extinct sea-scorpions), although only the Arachnida occur in amber. Arachnids are predators and feed on other invertebrates. Eight orders have been recorded in amber.

97 Baltic amber pendant containing a spider (Araneae) and harvestman (Opiliones). The harvestman (daddy-long-legs) is at the top. (Length of pendant excluding ring 50mm)

98 Woodlouse (Crustacea: Isopoda) in Dominican amber. (Length 2mm)

ARTHROPODS

The order Araneae (= Araneida) are the spiders with fang-like chelicerae and short, sometimes bulbous pedipalps. They usually have an unsegmented, fat hairy abdomen and hairy legs (fig 97). Spiders are common in amber and many different species have been described. Members of the order Opiliones are commonly known as harvestmen or 'daddy-long-legs'. They appear similar to spiders except the cephalothorax and abdomen are joined as one and they are not hairy (fig 97).

They also have long pedipalps and very long legs. Harvestmen are very rare in amber. The order Acari (= Acarida) are the mites and ticks. The cephalothorax and abdomen are joined as one, they usually have short legs and they are extremely small (often less than 1mm long) (fig 99). They can be smooth or hairy. Some juvenile mites only have six legs. Acarids are fairly common in amber but are often overlooked because of their small size. Mites are more common in Baltic amber than

Dominican amber. The order Scorpiones are the scorpions. Their pedipalps are modified to form pincers and they have a long tail with a sting (fig 100). Scorpions are extremely rare in amber, although they are often seen as fakes in plastic. The order Pseudoscorpionida are the pseudoscorpions. They superficially resemble scorpions in that they have pincer-like pedipalps, however they differ from scorpions in that they do not have a tail and they are much smaller (only a couple of millimetres long) (fig 101). They also have a fat segmented body and are very rare in amber. The order Amblypygi are commonly known as tailless whip scorpions. They are large arachnids with long spiny pedipalps and the front pair of legs are considerably longer than the others and are used as feelers. They are extremely rare in amber. The Solpugida are commonly known as wind spiders. They are large

99 Mite (Arachnida: Acari) in Baltic amber. (Length excluding legs 0.4mm)

100 Scorpion (Arachnida: Scorpiones) tail in Burmese amber. (Length 8.5mm)

101 Pseudoscorpion (Arachnida: Pseudoscorpionida) in Baltic amber. (Length including claws 3.2mm)

102 Schizomid (Arachnida: Schizomida) in Dominican amber. (Length including front legs 7.5mm)

and have four very large chelicerae; the two upper fangs point downwards and the two lower fangs point upwards. They act like pincers, with a similar action to a staple remover. The pedipalps are similar to the legs except they have flat adhesive pads at the end. Only one solpugid has been recorded from amber (Dominican). The Schizomida does not have a common name. Schizomids are small, blind arachnids (up to 6mm long) with a segmented cephalothorax and a short tail (fig 102). The chelicerae are large and fang-like. The front pair of legs are very long and used as feelers. They have only been found in Dominican amber.

3. MYRIAPODA – CENTIPEDES AND MILLIPEDES

Members of the superclass Myriapoda are long and thin with many body segments, many pairs of legs and a pair of antennae. There are four classes, three of which have been recorded from amber. The Chilopoda comprise the centipedes. They have a single pair of legs on each body segment and a pair of curved fangs situated under the head (fig 103). The Diplopoda comprise the millipedes. They have two pairs of legs on each body segment and generally have many more legs than centipedes (fig 104). They do not have fangs and, in contrast to centipedes, are vegetarians

rather than carnivores. One distinctive kind called *Polyxenus* is very hairy (fig 105). The Symphyla are similar to centipedes in that they have a single pair of legs on each segment, but they do not have fangs. Myriapods are very rare in amber. Centipedes are more common in Baltic amber than Dominican amber, although the reverse is true for millipedes.

4. HEXAPODA – INSECTS

The arthropods in the superclass Hexapoda have six legs and include the insects. They are the most abundant and diverse group of arthropods and will be described in more detail.

103 Centipede (Myriapoda: Chilopoda) in Baltic amber. (Width of centipede 1.1mm)

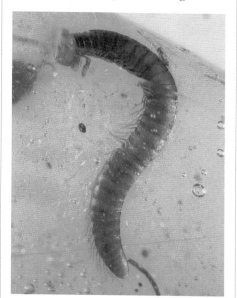

104 Millipede (Myriapoda: Diplopoda) in Dominican amber. (Width of millipede 0.6mm)

105 A very hairy millipede (Diplopoda: Polyxenus) in Baltic amber. (Length excluding hairs 4.5mm)

KEY TO ARTHROPODS

Keys to identification

These keys in this book are designed to identify most of the arthropod inclusions that have been found in amber and not living arthropods. They use the most obvious characteristics and are easy to follow. Many amber inclusions can only be examined from one side and parts of them are often obscured by cracks and bubbles. Pieces of amber may also be rounded, which can make the inclusion difficult to see due to distortion. If so, you may need to look at the photographs and descriptions to aid identification.

The rarity of an animal is indicated after its name by: VC very common, C common, R rare, VR very rare, (–) not recorded. Where this is given twice and separated by a slash, this indicates rarity in Baltic and then Dominican amber. For example, C/VR means common in Baltic amber but very rare in Dominican amber.

Key to arthropods

1 **Does it have wings or wingcases?**

YES
Insecta:
Pterygota
go to 28

NO
go to 2

2 **How many legs does it have?** (It's possible that legs have broken off)

6
Hexapoda
go to 14

8
(+ fangs
and/or
pedipalps)
Arachnida
go to 3

More than 8
go to 10

3 **Does it have a segmented abdomen?**

YES
go to 6

NO
go to 4

4 **Is there a constriction between the head and the abdomen?**

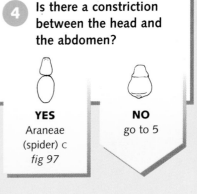

YES
Araneae
(spider) c
fig 97

NO
go to 5

KEY TO ARTHROPODS

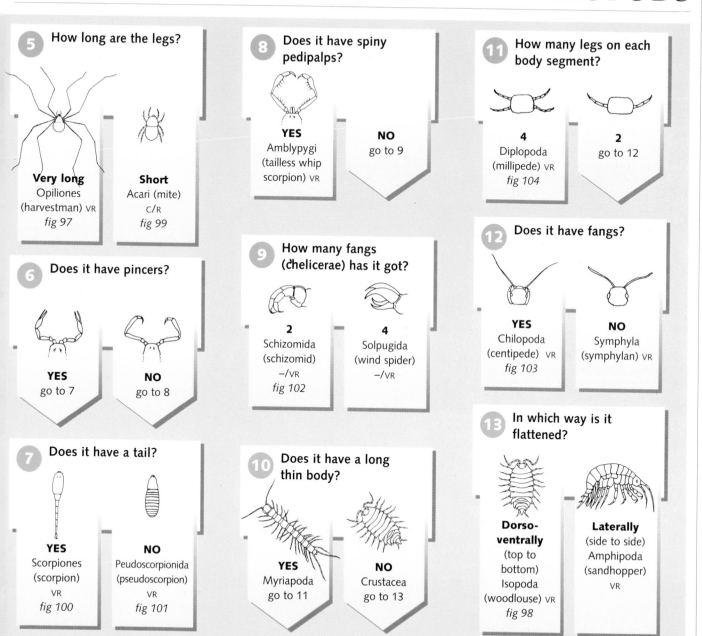

5 How long are the legs?

Very long
Opiliones
(harvestman) VR
fig 97

Short
Acari (mite)
C/R
fig 99

6 Does it have pincers?

YES
go to 7

NO
go to 8

7 Does it have a tail?

YES
Scorpiones
(scorpion)
VR
fig 100

NO
Pseudoscorpionida
(pseudoscorpion)
VR
fig 101

8 Does it have spiny pedipalps?

YES
Amblypygi
(tailless whip
scorpion) VR

NO
go to 9

9 How many fangs (chelicerae) has it got?

2
Schizomida
(schizomid)
–/VR
fig 102

4
Solpugida
(wind spider)
–/VR

10 Does it have a long thin body?

YES
Myriapoda
go to 11

NO
Crustacea
go to 13

11 How many legs on each body segment?

4
Diplopoda
(millipede) VR
fig 104

2
go to 12

12 Does it have fangs?

YES
Chilopoda
(centipede) VR
fig 103

NO
Symphyla
(symphylan) VR

13 In which way is it flattened?

Dorso-ventrally
(top to
bottom)
Isopoda
(woodlouse) VR
fig 98

Laterally
(side to side)
Amphipoda
(sandhopper)
VR

THE INSECT FOSSIL RECORD

Amber is extremely important for documenting the fossil record of terrestrial animals, particularly small insects that usually do not get preserved in sedimentary rocks. The amber insect record is complemented by insect-bearing lake sediments that preserve the larger insects such as dragonflies (Odonata) and grasshoppers (Orthoptera), which do not get trapped in amber. New discoveries are regularly being made in amber; for example there are some families of insects and other arthropods represented in the Dominican amber collection at The Natural History Museum, London, that have not previously been recorded in this amber. Some of these were not even known to have a fossil record.

106 *Graph showing the number of families of insects through time, according to the fossil record. The apparent jump from the Quaternary (Qua) to Recent (Rec) is a reflection of the number of living families without a fossil record.*

THE BALTIC AMBER FAUNA

The best studied amber insect fauna is that from the Baltic. About 5,000 species have been described and named so far, but this is by no means all of them. The total Baltic insect fauna about 40 million years ago may have been about 10,000 species. Some groups, although well represented in the amber, have hardly been studied, such as some families of parasitic wasps. Studies on the Dominican amber fauna only started about 30 years ago and so far only about 100 species have been described and named, although many more are currently being studied. The Dominican amber fauna is tropical and more diverse than that of the Baltic. Perhaps as many as 20,000 species were present in the Dominican Republic at the time. However, given that there are so many insect species that have not yet been named from amber or other fossil insect deposits there is no reliable database to determine the time either of first appearance, ie origination, or extinction of insect species.

The family is the most useful and reliable taxonomic unit to study insect origination and extinction. By the end of 1995 there were 1,275 families of insect in the fossil record, more are probably recorded now. This compares with 967 families living today, of which 648 have a fossil record (67%). Figure 106 illustrates the number of insect families at different periods of time, according to fossil evidence.

THE CRETACEOUS SURVIVORS

Amber has made the biggest contribution to the Upper Cretaceous, Eocene, Oligocene and Miocene insect record. It is interesting to note that 90% of insect families recorded from Canadian and Siberian amber (Upper Cretaceous) are represented by insects living today. This indicates that insects were hardly affected at the family level by the extinction at the end of the Cretaceous, which was when the dinosaurs became extinct. Unfortunately, there is no database to help us find out how many insect species became extinct at this time. A larger extinction of insect families took place during the Lower Cretaceous due to the radiation of the flowering plants, which changed terrestrial ecosystems.

The earliest records of extant insects come from Baltic amber. Eight species have been recorded but recent re-examination of four of them have shown that they are fake (fig 12), misidentified or preserved in copal. Of the other four, two are large, predominantly ground-dwelling beetles, so it is likely that these are also fakes. The other two records are as yet undisputed: they are a specimen of the mayfly *Heptagenia fuscogrisea* and several specimens of the tiny fairy fly *Palaeomymar duisburgi*, which is actually a parasitic wasp. Both belong to primitive groups that have hardly changed with time. Several species alive today are also known from Dominican and Mexican amber.

The superclass Hexapoda comprises four classes, all the members of which possess a three-segment thorax that has a single pair of legs on each segment and an abdomen with 12 or fewer segments. The first three classes, of which two have been recorded in amber, are primitive wingless hexapods (apterygotes) that have internal (entognathous) mouthparts. The Protura have not been found in amber.

COLLEMBOLA – SPRINGTAILS

The class Collembola are commonly known as springtails because they have a forked spring attached to the tip of the abdomen. At rest it is folded under the body (fig 107) but if the collembolan is disturbed, it flicks the fork backwards, which propels the animal into the air. Springtails are small (only a couple of millimetres long) and are fairly rare in amber, although they are often

overlooked because of their small size. Usually the forked spring can be seen projecting backwards, which is probably due to the animal trying to use it to escape from the amber. Springtails can be smooth or hairy and some have long antennae that curve outwards. One distinctive family recorded in amber is the Sminthuridae, which have a smooth, short, fat abdomen with a knob on the end (fig 108).

107 Springtail (Collembola) in Dominican amber. This example has not released its spring, which is still folded under its body. (Length 1.3mm)

108 Three springtails (Collembola: Sminthuridae) in Baltic amber. (Length of largest on the left, including spring 1.8mm)

109 Bristletail (Archaeognatha) in Baltic amber. (Length including tail 15mm)

110 Silverfish (Zygentoma) in Baltic amber. (Length 7mm)

111 A pair of overlapping damselfly wings in Baltic amber. (Length 17mm)

INSECTS

DIPLURA – TWO-TAILED BRISTLETAILS

The class Diplura are commonly known as two-tailed bristletails because they have two segmented appendages (cerci) at the tip of the abdomen. Unlike springtails, they do not have eyes and are blind. Diplurans are extremely rare in amber.

INSECTA – TRUE INSECTS

The majority of hexapods belong to the class Insecta, which are characterized by having external (ectognathous) mouthparts. The Insecta are split into two subclasses.

APTERYGOTA – WINGLESS INSECTS

The subclass Apterygota are the primitive insects that do not have wings and comprise two extant orders – the Archaeognatha and Zygentoma (also known as Thysanura). Both orders are similar in that they have long antennae and at least three long tails (caudal filaments) at the tip of the abdomen. They differ in that the Archaeognatha (known commonly as bristletails) have large eyes (fig 109). The Zygentoma (known commonly as silverfish), by contrast, have small eyes (fig 110). Both orders are rare in amber.

PTERYGOTA – WINGED INSECTS

The majority of insects belong to the subclass Pterygota, most of which possess two pairs of wings. This group also includes wingless groups (eg fleas) that evolved from winged ancestors. The wings have a network of veins that branch and are characteristic for each group of insect (see panel below). The wings are very useful for identifying the insects trapped in amber.

The Pterygota are divided into two smaller groupings – the Palaeoptera and Neoptera. The Palaeoptera comprise those insects that have outstretched wings whereas Neoptera are able to fold their wings back over their body.

Palaeoptera – outstretched wings

The Palaeoptera contains several orders, most of which are extinct. There are two living orders: the Odonata and Ephemeroptera. The Odonata are predators and have very short antennae,

GENERALIZED STRUCTURE OF A FLY'S WING

C: Costa. This vein forms the front edge (anterior) of the wing.
SC: Subcosta.
R: Radius. This usually branches posteriorly into the radial sector.
M: Median. This may branch.
Cu: Cubitus. This may also branch.
A: Anal vein(s).

 Pterostigma. This makes the wing stable during flight.

These main veins are connected by cross-veins and an area of the wing that is delimited by cross-veins is known as a cell. Key cross-veins and cells also have names.

112 Mayfly (Ephemeroptera) in Baltic amber. (Length excluding legs and tail 6.5mm)

very large eyes, powerful jaws, a long abdomen and equal-sized wings with many veins, cross-veins and a pterostigma. They comprise the dragonflies (suborder Anisoptera) and damselflies (suborder Zygoptera). The dragonflies are large and rest with their wings outstretched whereas damselflies are smaller and rest with their wings vertically. Dragonflies and damselflies are both extremely rare in amber, with dragonflies being rarer. The order Ephemeroptera comprise the mayflies. Mayflies have two or three long caudal filaments, short antennae and large compound eyes (fig 112). The wings have many cross-veins, do not possess a pterostigma and the hindwings are much smaller than the forewings.

Many mayfly families have been recorded in amber, although they are rare. This is due to the fact that they do not live long (a couple of hours to several days) and do not venture far from their pond or river breeding grounds.

Neoptera – folded wings
The majority of insects belong to the Neoptera, which consists of those insects that can fold their wings back along their body. The Neoptera can be divided into those that have incomplete metamorphosis (exopterygotes) and those that have complete metamorphosis (endopterygotes) (fig 114).

Incomplete metamorphosis
Exopterygotes start life as an egg, which hatches into a nymph generally looking similar to the adult, except that it is wingless. The nymph progressively grows by moulting and produces wings at its last moult to become an adult. (Palaeoptera are also exopterygotes).

Complete metamorphosis
Endopterygotes, by contrast, start life as an egg that hatches into a larva (also commonly known as a grub, maggot or caterpillar). The larva does not look anything like the adult and grows by progressively moulting. It then produces a pupa (chrysalis or cocoon) in which the tissue breaks down and re-forms (metamorphoses). The adult insect, complete with wings, then hatches out of the pupa.

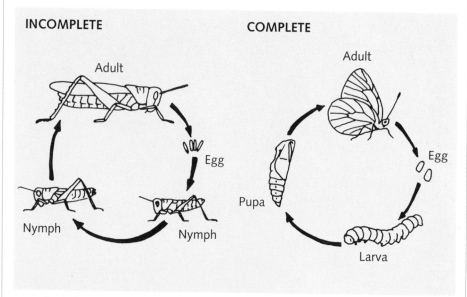

113 Neopteran: cockroach (Blattodea) in Baltic amber. (Length 12mm)

114 The differences between incomplete and complete metamorphosis.

KEY TO INSECTS WITHOUT WINGS

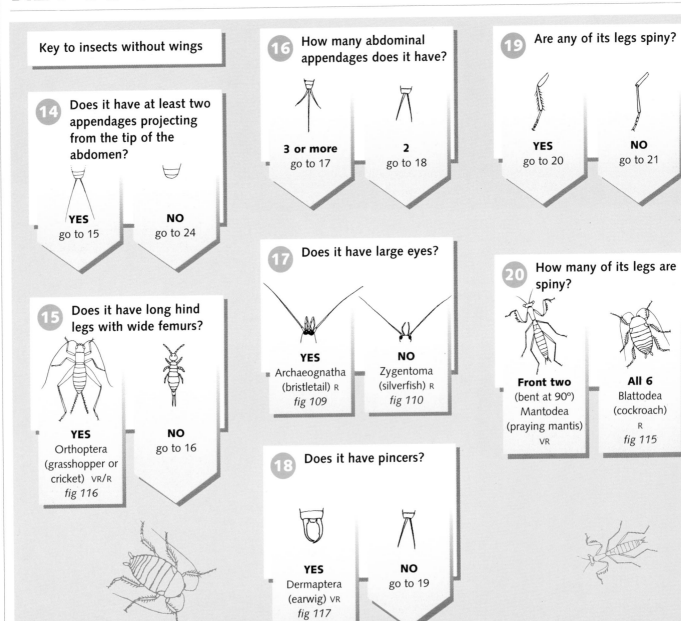

Key to insects without wings

14 Does it have at least two appendages projecting from the tip of the abdomen?

YES
go to 15

NO
go to 24

15 Does it have long hind legs with wide femurs?

YES
Orthoptera
(grasshopper or cricket) VR/R
fig 116

NO
go to 16

16 How many abdominal appendages does it have?

3 or more
go to 17

2
go to 18

17 Does it have large eyes?

YES
Archaeognatha
(bristletail) R
fig 109

NO
Zygentoma
(silverfish) R
fig 110

18 Does it have pincers?

YES
Dermaptera
(earwig) VR
fig 117

NO
go to 19

19 Are any of its legs spiny?

YES
go to 20

NO
go to 21

20 How many of its legs are spiny?

Front two
(bent at 90°)
Mantodea
(praying mantis)
VR

All 6
Blattodea
(cockroach)
R
fig 115

KEY TO INSECTS WITHOUT WINGS

23 Does it have eyes?

YES
Collembola
(springtail)
R
fig 107

NO
Diplura
(two-tailed
bristletail)
VR

26 Is the body laterally compressed (flattened from side to side)?

YES
Siphonaptera
(flea) VR

NO
go to 27

21 Are the tips of its forelegs enlarged?

YES
Embioptera
(web spinner)
VR

NO
go to 22

24 Is there a strong constriction between the thorax and abdomen?

YES
Hymenoptera:
Formicidae
(ant) C/VC
fig 150

NO
go to 25

27 Does it have a proboscis?

YES
Hemiptera
(bug)
go to 52

NO
Probably
another
wingless insect
(adult, nymph
or larva)

22 Does it have a long thin body with elongated thorax?

YES
Phasmatodea
(stick insect) VR
fig 118

NO
go to 23

25 Does it have a narrow thorax?

YES
probably
Isoptera
(termite) R/C
fig 96

NO
go to 26

INCOMPLETE METAMORPHOSIS

There are many orders of insects with incomplete metamorphosis in their life cycle. Some of the orders are extinct, but all of those that occur in amber are extant.

BLATTODEA – COCKROACHES

Cockroaches are dorso-ventrally flattened insects with toughened forewings (tegmina) and a head-shield (pronotum) (fig 113). They have very long antennae, very spiny legs and cerci that are usually hairy. The wings have a dense venation with many parallel and branching veins with few or no cross-veins. The radius sends many parallel branches towards the costa, and the anal veins occur in a distinctive curved or triangular area at the base of the wing known as the clavus. Cockroaches are fairly rare in amber because they are large and strong and can therefore pull themselves free of the resin. Both adults and nymphs occur in amber, but the nymphs do not have wings (fig 115).

ISOPTERA – TERMITES

Termites have small eyes and very short cerci that are very difficult to see. They are social insects with winged, worker and soldier castes (fig 96). The winged castes have membranous wings with a wrinkled or knobbly texture and faint, unpigmented veins. The costa, subcosta and radius are the most distinct veins and run parallel to each other. The workers are wingless and generally have short, fat, segmented bodies. The head is usually wider than the thorax. Termites are common in Dominican amber but rare in Baltic amber and they often have large bubbles projecting from their bodies. These bubbles consist of methane produced by the many bacteria in the gut that helped the termite to digest its food.

ORTHOPTERA – GRASSHOPPERS, CRICKETS AND LOCUSTS

Orthoptera possess hind legs that have wide femurs and are much longer than the other legs. These adaptations enable them to jump. Orthoptera are rare in amber; crickets (superfamily Grylloidea) and several kinds of grasshopper have been recorded. Crickets have short, fat bodies with long, hairy cerci and long antennae (fig 116). They sometimes have wings that lie flat over the body. In males the wing veins are convoluted to enable singing, whereas in females the veins are straight and have many cross-veins. By contrast, grasshoppers generally have longer bodies, short cerci and their wings are held vertically along their bodies. Some female Orthoptera have a long, curved, pointed ovipositor projecting from the abdomen, which is used for laying eggs.

115 Cockroach nymph (Blattodea) in Baltic amber. (Length 3.4mm)

116 Cricket (Orthoptera: Grylloidea) in Dominican amber. (Length 5.7mm)

117 Earwig (Dermaptera) in Dominican amber. (Length 7mm)

INCOMPLETE METAMORPHOSIS

DERMAPTERA – EARWIGS

Earwigs have elongate bodies with cerci that are modified to form pincers (fig 117). They have short wing-cases (elytra) with a fan-like hindwing folded underneath. The antennae are generally long. Earwigs are extremely rare in amber, which is surprising considering they are commonly found in the cracks in the bark of trees today.

EMBIOPTERA – WEB SPINNERS

Web spinners are small (a few millimetres long) with an elongate body, small asymmetrical cerci and the ends of their front legs are often greatly enlarged. Some have wings, others are wingless. The winged forms have two pairs of equal-sized wings with a few longitudinal veins. They are a tropical group that is extremely rare in amber.

MANTODEA – PRAYING MANTISES

Praying mantises have elongate bodies and a wide head with large eyes and long antennae. Their front legs are held upwards and bent by at least 90 degrees. They are used to catch prey and often have many spines. Mantids are extremely rare in amber.

PHASMATODEA – STICK OR LEAF INSECTS

Phasmatodeans are either very long and thin (often wingless) to mimic twigs or flat and wide to mimic leaves. They are very rare in amber, and only the long and thin kind have been found (fig 118).

PLECOPTERA – STONEFLIES

Stoneflies have long antennae, long cerci and two pairs of equal-sized wings. The venation is distinctive in that there are many cross-veins between the median and cubital veins, giving a ladder-like appearance (fig 119). They are very rare in amber. Most of those in Baltic amber belong to the family Leuctridae, which wrap their wings around their body when at rest.

PSOCOPTERA – BARKLICE AND BOOKLICE

Psocopterans are very small (a couple of millimetres long) with two pairs of wings held roof-like over the short body and a squarish head that has long, thin antennae. The forewings are slightly larger than the hindwings and often have a large pterostigma, or they can be pointed and hairy (fig 120). They are fairly common in Dominican amber but very rare in Baltic amber, and many families have been recorded.

118 *Stick insect (Phasmatodea) in Baltic amber. (Length 11mm)*

119 *Stonefly (Plecoptera: Leuctridae) in Baltic amber. (Length 8mm)*

120 *Barklouse (Psocoptera) in Dominican amber. (Length 1.6mm)*

INCOMPLETE METAMORPHOSIS

THYSANOPTERA – THRIPS

Thrips are very small (usually less than 2mm long), thin and the adults have two pairs of wings. The wings are of equal size, narrow with one or two veins and have a distinctive fringe of hairs (fig 121). They are rare in amber but have probably been overlooked because of their small size.

ZORAPTERA

The order Zoraptera are small, slightly hairy insects with an elongate thorax. They are either apterous (wingless) or have two pairs of wings with a few veins. The apterous ones do not have any eyes and are blind. Only one zorapteran has been recorded from amber (Dominican).

PHTHIRAPTERA – LICE

Phthirapterans are apterous, flattened, parasitic insects with clawed feet. The only record of this order in amber is of eggs on mammal hair in Baltic amber.

121 Thrip (Thysanoptera) in Dominican amber. (Length 0.7mm)

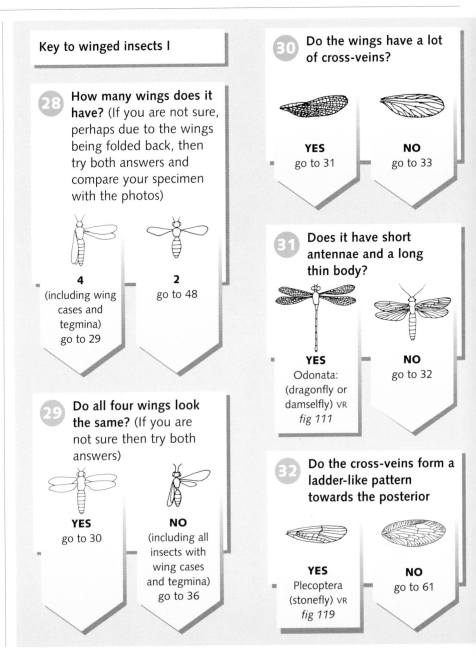

Key to winged insects I

28 How many wings does it have? (If you are not sure, perhaps due to the wings being folded back, then try both answers and compare your specimen with the photos)

4
(including wing cases and tegmina)
go to 29

2
go to 48

29 Do all four wings look the same? (If you are not sure then try both answers)

YES
go to 30

NO
(including all insects with wing cases and tegmina)
go to 36

30 Do the wings have a lot of cross-veins?

YES
go to 31

NO
go to 33

31 Does it have short antennae and a long thin body?

YES
Odonata:
(dragonfly or damselfly) VR
fig 111

NO
go to 32

32 Do the cross-veins form a ladder-like pattern towards the posterior

YES
Plecoptera (stonefly) VR
fig 119

NO
go to 61

33 Are the wings scaly, hairy or neither?

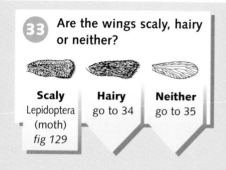

Scaly
Lepidoptera
(moth)
fig 129

Hairy
go to 34

Neither
go to 35

34 Is it small (body length less than 2mm)?

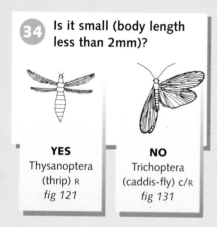

YES
Thysanoptera
(thrip) R
fig 121

NO
Trichoptera
(caddis-fly) C/R
fig 131

35 Are the ends of its forelegs enlarged?

YES
Embioptera
(web spinner)
VR

NO
Isoptera
(termite)
R/C
fig 96

36 Are the hind wings smaller than the forewings?
(If you are not sure then try both answers)

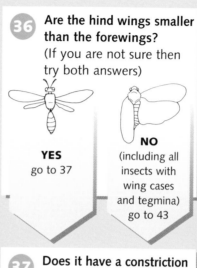

YES
go to 37

NO
(including all insects with wing cases and tegmina)
go to 43

37 Does it have a constriction between the thorax and abdomen?

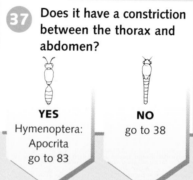

YES
Hymenoptera:
Apocrita
go to 83

NO
go to 38

38 Does it have at least two long tails (caudal filaments)?

YES
Ephemeroptera
(mayfly) R
fig 112

NO
go to 39

39 Does it have lots of cells in the wing?

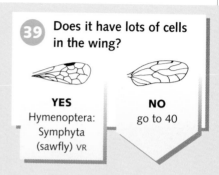

YES
Hymenoptera:
Symphyta
(sawfly) VR

NO
go to 40

40 Is the body hairy?

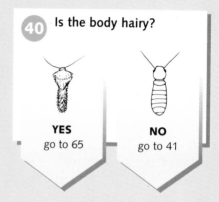

YES
go to 65

NO
go to 41

41 Does only one vein originate from the base of the wing?

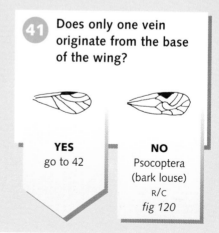

YES
go to 42

NO
Psocoptera
(bark louse)
R/C
fig 120

KEY TO WINGED INSECTS I

42 Is the vein unbranched?

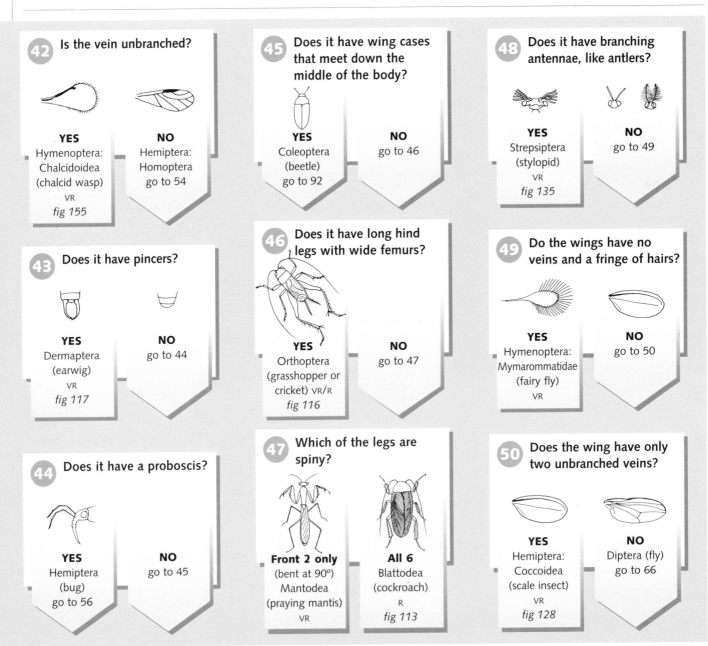

YES
Hymenoptera:
Chalcidoidea
(chalcid wasp)
VR
fig 155

NO
Hemiptera:
Homoptera
go to 54

43 Does it have pincers?

YES
Dermaptera
(earwig)
VR
fig 117

NO
go to 44

44 Does it have a proboscis?

YES
Hemiptera
(bug)
go to 56

NO
go to 45

45 Does it have wing cases that meet down the middle of the body?

YES
Coleoptera
(beetle)
go to 92

NO
go to 46

46 Does it have long hind legs with wide femurs?

YES
Orthoptera
(grasshopper or cricket) VR/R
fig 116

NO
go to 47

47 Which of the legs are spiny?

Front 2 only
(bent at 90°)
Mantodea
(praying mantis)
VR

All 6
Blattodea
(cockroach)
R
fig 113

48 Does it have branching antennae, like antlers?

YES
Strepsiptera
(stylopid)
VR
fig 135

NO
go to 49

49 Do the wings have no veins and a fringe of hairs?

YES
Hymenoptera:
Mymarommatidae
(fairy fly)
VR

NO
go to 50

50 Does the wing have only two unbranched veins?

YES
Hemiptera:
Coccoidea
(scale insect)
VR
fig 128

NO
Diptera (fly)
go to 66

122 Bug (Hemiptera: Heteroptera) in Dominican amber. (Length 1.7mm)

123 Planthopper (Homoptera: Fulgoroidea) in Baltic amber. (Length of forewing 6.5mm)

124 Froghopper (Homoptera: Cercopoidea) in Baltic amber. (Length 6.5mm)

HEMIPTERA – BUGS

The order Hemiptera comprise the bugs. They have a tube-like proboscis that they use to suck plant sap or blood. There are two suborders: the Heteroptera and Homoptera, although some scientists use the word bug for the Heteroptera only.

HETEROPTERA

The Heteroptera have a curved proboscis that projects downwards from the front of the head and long antennae consisting of only a few segments. They also have forewings that are usually divided into two parts. The basal half of the wing is thickened (sclerotized) whereas the distal half is membranous. The membranous parts of the forewings overlap when the wings are folded (fig 122). Heteroptera are fairly rare in amber although many families have been recorded. The families generally appear similar to each other and are difficult to identify. Members of a

125 Leafhopper (Homoptera: Cicadellidae) in Dominican amber. (Length 3mm)

distinctive subfamily of assassin bugs (Reduviidae: Emesinae) have an elongate abdomen and very long legs.

HOMOPTERA

Members of the suborder Homoptera have a straight proboscis that originates at the back of the head and projects under the body. The forewings are not separated into two parts. Homoptera are fairly common in amber and the different superfamilies are fairly easy to distinguish.

The Fulgoroidea, Cicadelloidea and Cercopoidea are similar in that their forewings are sclerotized and held roof-like over the body. They have extremely short antennae that are barely visible. The forewings have a distinctive triangular clavus and are often patterned. The Fulgoroidea (commonly known as planthoppers) are large bugs with three distinctive ridges running down the front of the head, a few spines on the legs and the two anal veins in the clavus merge to form a Y-shape (fig 123). They are fairly rare in amber. The Cicadelloidea (commonly known as leaf or treehoppers) have a triangular head and two anal veins in the clavus that do not merge. The family Cicadellidae (commonly known as leafhoppers) have spiny hindlegs and are fairly common in Dominican amber but rare in Baltic amber (fig 125). The family Membracidae (commonly known as treehoppers) have large head shields of various shapes and sizes, and are known

BUGS

from Dominican amber, where they are extremely rare. The superfamily Cercopoidea (commonly known as froghoppers or spittlebugs) do not have any veins visible in the clavus and are extremely rare in amber (fig 124).

The superfamilies Aphidoidea, Psylloidea, Aleyrodoidea and Coccoidea are small insects (generally less than 3mm long) with membranous wings that only have a few veins. The Aphidoidea are known commonly as aphids or greenfly. They have fat bodies and sometimes hold their wings vertically when at rest. The forewings are much larger than the hindwings, have a pterostigma and a prominent vein running close to the anterior edge of the wing with a few veins coming obliquely off it (fig 126). They often have tubes on the body, which would have ejected blood to repel predators (fig 127) and they often have an extremely long proboscis that can be

twice the length of the body. They are fairly common in Baltic amber but extremely rare in Domincan amber. The Psylloidea (commonly known as jumping plant lice) have forewings that are slightly larger than the hindwings and possess short antennae. The forewings have a single main vein that branches a few times. They are extremely rare in amber. The Aleyrodoidea are known commonly as whiteflies because their wings are white. They have two pairs of wings with a single vein that forks near the middle of the wing. They are extremely rare in amber. The Coccoidea are known commonly as scale insects. Only the males have wings, of which only a single pair is present. The wings usually have two unbranched veins: the first vein runs close to the anterior margin, the second runs through the middle of the wing (fig128). The wing often has a wrinkled texture. Scale insects are very rare in amber.

126 Winged aphid (Homoptera: Aphidoidea) in Baltic amber. (Length 3.5mm)

127 Aphid (Homoptera: Aphidoidea) in Baltic amber. (Length 1.7mm)

128 Scale insect (Homoptera: Coccoidea) in Baltic amber. (Length of wing 1.2mm)

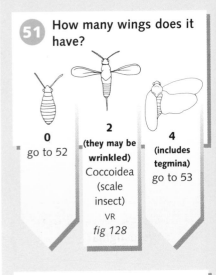

Key to bugs

51 How many wings does it have?

0
go to 52

2
(they may be wrinkled)
Coccoidea (scale insect)
VR
fig 128

4
(includes tegmina)
go to 53

52 Is its proboscis longer than its body?

YES
Aphidoidea (greenfly) C/VR
fig 127

NO
go to 56

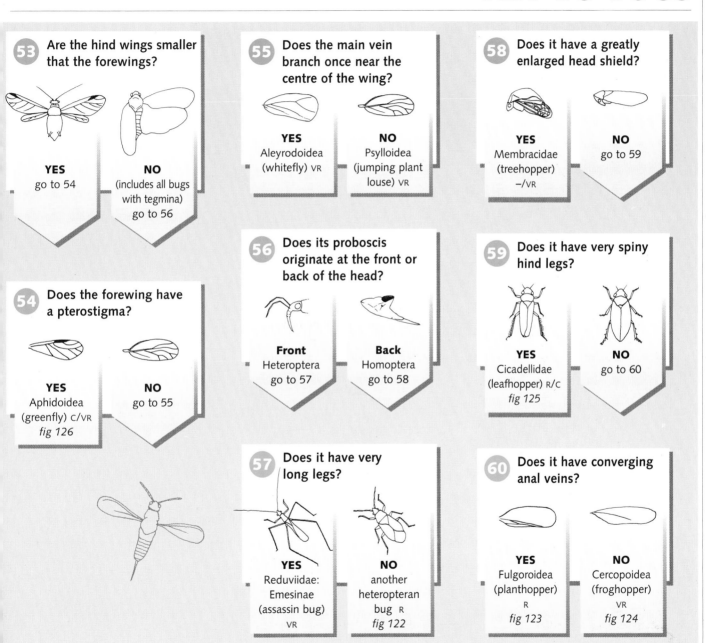

53 Are the hind wings smaller that the forewings?

YES
go to 54

NO
(includes all bugs with tegmina)
go to 56

54 Does the forewing have a pterostigma?

YES
Aphidoidea
(greenfly) C/VR
fig 126

NO
go to 55

55 Does the main vein branch once near the centre of the wing?

YES
Aleyrodoidea
(whitefly) VR

NO
Psylloidea
(jumping plant louse) VR

56 Does its proboscis originate at the front or back of the head?

Front
Heteroptera
go to 57

Back
Homoptera
go to 58

57 Does it have very long legs?

YES
Reduviidae:
Emesinae
(assassin bug)
VR

NO
another
heteropteran
bug R
fig 122

58 Does it have a greatly enlarged head shield?

YES
Membracidae
(treehopper)
–/VR

NO
go to 59

59 Does it have very spiny hind legs?

YES
Cicadellidae
(leafhopper) R/C
fig 125

NO
go to 60

60 Does it have converging anal veins?

YES
Fulgoroidea
(planthopper)
R
fig 123

NO
Cercopoidea
(froghopper)
VR
fig 124

COMPLETE METAMORPHOSIS

129 Moth (Lepidoptera) in Baltic amber. (Length 5.5mm)

The endopterygotes are those insects with complete metamorphosis. There are 11 extant orders, all of which have been found in amber.

LEPIDOPTERA – MOTHS AND BUTTERFLIES

Moths and butterflies usually have two pairs of scaly wings, a long, curled proboscis and a hairy body. Moths hold their wings back along their body whereas butterflies are generally larger and hold their wings vertically above the body when at rest. Moths are fairly rare in amber and butterflies are extremely rare, although the latter are sometimes seen as fakes. The scales on the wings refract the light to produce the bright colour patterns. However, the amber penetrates the wings and original colour patterns are lost or very faded (fig 129). (Some other insects, eg planthopper bugs (fig 61), do have colour patterns preserved but this is because the wings are pigmented, which is not affected by the amber.) It is common to find many of the scales embedded in the surrounding amber, which would have detached while the moth was struggling to escape. Sometimes the proboscis is partially uncurled in the amber. Many different families of moths have been recorded in amber, but they are often difficult to identify. Caterpillars also occur in amber (fig 130).

TRICHOPTERA – CADDIS-FLIES

Caddis-flies have two pairs of wings, which they hold roof-like over the body.

130 Caterpillar (Lepidoptera) in Baltic amber. (Length 8.5mm)

131 Caddis-fly (Trichoptera) in Baltic amber. (Length 8mm)

132 Scorpion fly (Mecoptera) in Baltic amber. (Length of wing 11mm)

Caddis-flies are very similiar in appearance to moths except they have hairy wings rather than scaly, and they have chewing mouthparts rather than a proboscis (fig 131). They are common in Baltic amber but fairly rare in Dominican amber. Many different families have been recorded but they are difficult to tell apart.

MECOPTERA – SCORPION FLIES

Scorpion flies have two pairs of equal-sized wings with veins that progressively branch and have cross-veins between them (fig 132). Scorpion flies have a characteristic elongate head that points downwards, long antennae and the males often have a long tail that curves upwards. They are extremely rare in Baltic amber only and three families have been recorded.

NEUROPTERA – LACEWINGS AND ANT-LIONS

Lacewings and ant-lions have two pairs of equal-sized wings with many branching veins and cross-veins. Most of the veins branch immediately before they reach the margin (fig 133). They are very rare in amber but several families have been recorded. Larvae with large jaws are also known (fig 134). One distinctive family (Mantispidae), known as mantis flies, have long, bent spiny forelegs similar to those of a praying mantis. It is surprising that Neuroptera are rare in Baltic amber because some families feed on aphids, which are common in Baltic amber.

MEGALOPTERA – ALDER FLIES

Alder flies appear similar to lacewings except they have much thicker wing veins (fig 60). They are extremely rare in Baltic amber only.

133 Lacewing (Neuroptera) in Baltic amber. (Length 8mm)

134 Lacewing larva (Neuroptera) in Baltic amber. (Length including jaws 3mm)

COMPLETE METAMORPHOSIS

RAPHIDIOPTERA – SNAKE FLIES

Snake flies also appear similar to lacewings except they have an elongated thorax. Like alder flies they are extremely rare in Baltic amber and have not been recorded in Dominican amber.

SIPHONAPTERA – FLEAS

Fleas are specialized vertebrate parasites. They are very small (1mm or less), wingless and have a laterally compressed body, which is hairy. They are extremely rare in amber.

STREPSIPTERA – STYLOPIDS

Stylopids are very small (typically 1mm long) with a single pair of fan-like wings and branching antennae that look like antlers (fig 135). They are specialized parasites of other insects and extremely rare in amber.

Figure 135 Stylopid (Strepsiptera) in Dominican amber. (Length 1.5mm)

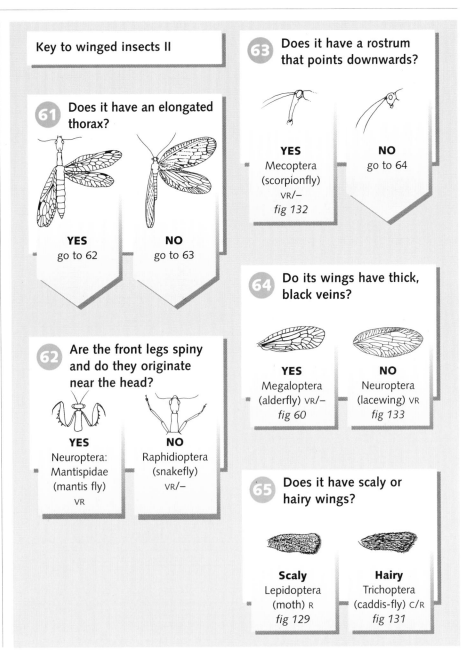

Key to winged insects II

61 Does it have an elongated thorax?

YES
go to 62

NO
go to 63

62 Are the front legs spiny and do they originate near the head?

YES
Neuroptera:
Mantispidae
(mantis fly)
VR

NO
Raphidioptera
(snakefly)
VR/–

63 Does it have a rostrum that points downwards?

YES
Mecoptera
(scorpionfly)
VR/–
fig 132

NO
go to 64

64 Do its wings have thick, black veins?

YES
Megaloptera
(alderfly) VR/–
fig 60

NO
Neuroptera
(lacewing) VR
fig 133

65 Does it have scaly or hairy wings?

Scaly
Lepidoptera
(moth) R
fig 129

Hairy
Trichoptera
(caddis-fly) C/R
fig 131

COMPLETE METAMORPHOSIS – FLIES

DIPTERA – TRUE FLIES

The order Diptera are the flies, which are characterized by only having a single pair of wings. The hindwings are reduced to small club-like structures known as halteres, which are used as balancers during flight. Flies are the most abundant insects in Baltic amber and even swarms are preserved (fig 1).

There are three suborders of Diptera. The Nematocera are easy to identify in that they have long, multi-segmented antennae. The Brachycera and Cyclorrhapha (= Muscomorpha) have short antennae, often with a bristle known as an arista.

136 Fungus gnat (Nematocera: Mycetophilidae) in Baltic amber. This specimen is a male, as shown by its external genitalia. (Length 5.5mm)

1. NEMATOCERA

The Nematocera are the most primitive suborder of flies and are the most common insects in Baltic amber. The males have external genitalia consisting of a pair of hooks. Many families have been recorded and those most likely to be encountered are briefly described here.

The Tipulidae are commonly known as crane-flies or daddy-long-legs (the latter is also used for harvestmen spiders). They are easy to identify in that they are generally large flies with extremely long legs, a long slender body, an elongate head and slender wings (fig

137 Fungus gnat (Nematocera: Sciaridae) in Baltic amber. This specimen is a female, as shown by its tapered abdomen. (Length 3mm)

62). Often, legs have broken away from the body due to the insect struggling. The wings have many veins and frequently have a squarish cell towards the tip of the wing. They are fairly rare in amber.

The Mycetophilidae are known commonly as fungus gnats. They generally have long antennae and rounded wings (fig 136). Their bodies are often long and slender, but they can be short and fat. This family is very diverse and many different subfamilies and genera have been recorded. The wing venation is distinct and both the median and cubital veins branch once, usually in the centre of the wing. The radius and radial sector are connected by one or two cross-veins. The radial sector terminates at the tip of the wing and sometimes sends a branch to the anterior margin, near the tip. They are fairly common in amber. The Sciaridae are a closely related family of fungus gnats. They differ from mycetophilids in that only the median vein branches in the centre of the wing and those branches are strongly curved (fig 137). The cubitus branches near the base of the wing. There is only one cross-vein between the radius and radial sector, and the radial sector does not branch. Sciarids are very common in Baltic amber but rare in Dominican amber.

The Scatopsidae are the scavenger flies. They have short bodies, fairly short antennae and rounded wings (fig 63). The venation is similar to that of

FLIES

sciarids in that only the median branches in the centre of the wing and those branches are curved. The venation differs from sciarids in that the radius and radial sector curve upwards and terminate near the centre of the anterior margin of the wing. They are fairly common in Dominican amber but very rare in Baltic amber.

The Chironomidae, Ceratopogonidae and Cecidomyiidae are the midges, which are generally only a couple of millimetres long. The wing venation is useful in telling them apart, but because of their small size this isn't easy to see without a microscope.

The chironomids are the non-biting midges, which have slender bodies and slender wings (fig 138). The males have extremely feathery antennae. The radius is straight, running parallel to the anterior margin, and is connected to the radial sector by a single cross-vein. The radial sector is straight and terminates near the tip of the wing. The median and cubitus veins are faint. The cubitus forks once but the median does not branch.

The ceratopogonids are the biting midges. They differ from chironomids in that the body is shorter and fatter and the wings are rounded (fig 92). They also have elongate mouthparts projecting downwards. The radius and radial sector are connected by two cross-veins and they curve upwards and terminate near the centre of the anterior margin. Both the median and cubitus veins fork once.

The cecidomyiids are the gall midges. They are similar to ceratopogonids in that they have fat bodies and rounded wings, however they have much longer antennae with bead-like segments (fig 139). The wing venation is similar to that of chironomids in that the radial sector terminates at the tip of the wing. The median is often absent; if present it does not branch. Usually cecidomyiid wings are hairy, however some species of chironomids and ceratopogonids also have hairy wings. Chironomids are very common in Baltic amber, but rare in Dominican amber. Ceratopogonids are fairly common in amber. Cecidomyiids are rare in Baltic amber, but fairly

138 Pair of mating midges (Nematocera: Chironomidae) in Baltic amber with two air bubbles. (Length of male on right with feathery antennae 1.6mm)

139 Gall midge (Nematocera: Cecidomyiidae) in Baltic amber (Length 1.8mm)

common in Dominican amber.

The Psychodidae are commonly known as moth flies, owl-midges or sandflies. They are small, hairy flies with bead-like antennal segments (fig 140). The wings are generally hairy with many equally spaced, parallel veins. The subfamily Phlebotominae are the bloodsucking sandflies, which have elongate mouthparts (fig 91). Psychodids are rare in Baltic amber, but fairly common in Dominican amber.

Apart from the biting midges and sandflies, there are two other bloodsucking nematoceran families. They are the Culicidae (mosquitoes) and Simuliidae (black-flies). Mosquitoes generally appear similar to chironomid midges in that they have long slender bodies and slender wings (fig 90). They differ from chironomids in that

mosquitoes are larger, with a long thin proboscis and scaly wings. The wing venation differs in that the radial sector and median veins branch. Many people think that mosquitoes are common in amber, however this is not the case. Only a few are known in Baltic amber and a few tens of specimens are known in Dominican amber. Most so-called 'mosquitoes' in amber actually belong to other groups; most turn out to be fungus gnats. The black-flies – not to be confused with blackfly (Hemiptera: Aphidoidea) – have short, fat bodies and short antennae (fig 94). The wings appear similar to those of mycetophilids in that they are rounded. They also have a radial sector that terminates at the wing tip and a median vein that forks. Simuliids differ from mycetophilids in that the median forks very close to the

point where the radial sector is connected to the radius (by a cross-vein) rather than in the centre of the wing (fig 94). The cubital veins originate right at the base of the wing and one of them is often kinked. Simuliids are very rare in Baltic amber.

2. BRACHYCERA

The suborders Brachycera and Cyclorrhapha are flies that usually have very short antennae. The Brachycera generally have slender antennae consisting of a few segments with the hair-like arista originating from the tip.

Many families of Brachycera have been recorded in amber but only three are regularly encountered. The Rhagionidae are commonly known as snipe flies. They are large, robust flies with rounded wings that usually possess

140 Moth fly in Baltic amber (Nematocera: Psychodidae) (Length 1.3mm)

141 Snipe flies (Brachycera: Rhagionidae) in Baltic amber. (Length of top fly 5mm)

142 Dance fly (Brachycera: Empididae) in Baltic amber. (Length 3.7mm)

FLIES

143 Long-legged fly (Brachycera: Dolichopodidae) in Baltic amber. (Length 2.3mm)

144 Advanced fly (Cyclorrapha) in Baltic amber. (Length 4.3mm)

145 Scuttle fly (Cyclorrapha: Phoridae) in Dominican amber. (Length 1.7mm)

a pterostigma (fig 141). The wing has many veins and a distinctive medial cell with three or four veins projecting from it that terminate on the posterior margin. Rhagionids are fairly rare in Baltic amber and extremely rare in Dominican amber.

There are several families that have been recorded from amber that are closely related to rhagionids, however they are extremely rare. Of these the Rachiceridae differ from rhagionids in having long multi-segmented antennae (fig 75). The Tabanidae are the bloodsucking horseflies. They are larger than rhagionids and have a widely diverging forked radial sector at the tip of the wing (fig 93).

The Empididae are commonly known as dance flies. They are generally smaller than rhagionids and often have a long pointed proboscis projecting down from the head (fig 142). The wing venation is diverse in this group. The radial sector terminates at the tip of the wing and is either unbranched or forks near the tip. There are several subfamilies that have an elongate medial cell with two or three veins projecting from it. Most also have a cubital vein that does not reach the posterior margin, instead two cross-veins project horizontally from it to produce an oblique, upside-down T-shape.

Some species have neither the medial cell or T-shaped cubitus and these belong to the subfamily Tachydromiinae. They have a simpler venation with an unbranched median that produces two cross-veins. One cross-vein connects to the radial sector and the other to the cubitus. The cross-veins either originate at the same point near the centre of the wing or are slightly staggered. Empidids are fairly rare in amber.

The Dolichopodidae are small flies that are known as long-legged flies, although their legs are not noticably longer than those of other groups. In the males the external genitalia are large and are often curved under the body. The wings have a simple venation with a few unbranched veins (fig 143). There is a single, distinctive cross-vein between the median and cubitus situated towards the posterior margin. Sometimes the median is kinked before it terminates at the tip of the wing. Dolichopodids are very common in Baltic amber but fairly rare in Dominican amber.

3. CYCLORRHAPHA

The Cyclorrhapha are the most advanced suborder of flies and are the most common group living today, however they are rare in amber. They generally have blob-like antennae with the arista originating near the base. Sometimes the arista of cyclorrhaphans is hairy. Many families have been recorded but they are very difficult to tell apart by a non-expert. Most have a simple venation very similar to that of dolichopodids except they also have a cross-vein between the median and radial sector near the centre of the wing (fig 144). One such family is the Drosophilidae (fruit flies), which is only known in the fossil record from amber (fig 74). A recognisable family is the Phoridae, which are commonly known as scuttle flies. They are small, black, hairy flies that have spiny palps (sensory appendages next to the mouth) and rounded wings with a distinctive venation (fig 145). The radius and radial sector are thick and terminate on the anterior margin. The four other veins originate from the radial sector or the base of the wing. They are unbranched, equally spaced and curve slightly to terminate at the tip of the wing or on the posterior margin. Phorids are fairly rare in amber. Another distinctive family are the Syrphidae (hoverflies). These are large flies that are generally not hairy. The median and cubitus veins do not reach the posterior margin; instead, they bend upwards and join the vein above (fig 146). They are rare in Baltic amber and extremely rare in Dominican amber.

146 Hoverfly (Cyclorrapha: Syrphidae) in Baltic amber. (Length 6mm)

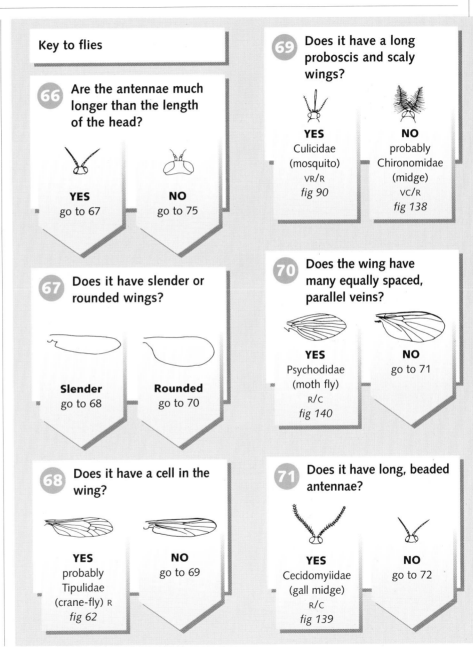

Key to flies

66 Are the antennae much longer than the length of the head?

YES
go to 67

NO
go to 75

67 Does it have slender or rounded wings?

Slender
go to 68

Rounded
go to 70

68 Does it have a cell in the wing?

YES
probably Tipulidae (crane-fly) R
fig 62

NO
go to 69

69 Does it have a long proboscis and scaly wings?

YES
Culicidae (mosquito)
VR/R
fig 90

NO
probably Chironomidae (midge)
VC/R
fig 138

70 Does the wing have many equally spaced, parallel veins?

YES
Psychodidae (moth fly)
R/C
fig 140

NO
go to 71

71 Does it have long, beaded antennae?

YES
Cecidomyiidae (gall midge)
R/C
fig 139

NO
go to 72

KEY TO FLIES

72 Does the radial sector reach the wing tip?

YES
go to 73

NO
go to 74

---

75 Does the wing have any cells?

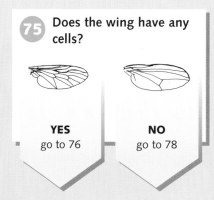

YES
go to 76

NO
go to 78

---

78 How many cross-veins does it have?

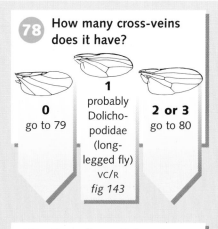

0
go to 79

1
probably Dolicho-podidae (long-legged fly)
VC/R
fig 143

2 or 3
go to 80

---

73 Do the cubitus or radial sector branch?

YES
probably Mycetophilidae (fungus gnat)
C
fig 136

NO
probably Sciaridae (fungus gnat)
VC/R
fig 137

---

76 Does it have veins running close to the posterior margin?

YES
probably Syrphidae (hover fly)
R/VR
fig 146

NO
go to 77

---

79 Does the radial sector reach the tip of the wing?

YES
Simuliidae (black fly)
VR/–
fig 94

NO
Phoridae (scuttle fly)
R
fig 145

---

74 Does the cubitus branch?

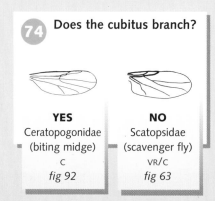

YES
Ceratopogonidae (biting midge)
C
fig 92

NO
Scatopsidae (scavenger fly)
VR/C
fig 63

---

77 Does the cubitus converge with the anal vein?

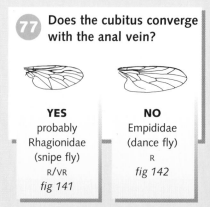

YES
probably Rhagionidae (snipe fly)
R/VR
fig 141

NO
Empididae (dance fly)
R
fig 142

---

80 Are the cross-veins together near the centre of the wing?

YES
Empididae: Tachydromiinae (dance fly)
R

NO
probably Cyclorrapha (fly)
R
fig 144

HYMENOPTERA – SAWFLIES, WASPS, BEES AND ANTS

Insects in the order Hymenoptera have two pairs of wings with a reduced venation. The forewings usually have a pterostigma and are larger than the hindwings. There are two suborders: the Symphyta and the Apocrita.

SYMPHYTA – SAWFLIES

The Symphyta are large and have many distinct cells in their wings. Sawflies are extremely rare in amber.

APOCRITA – WASPS, BEES AND ANTS

The Apocrita are the wasps, bees and ants. They differ from the Symphyta in that there is usually a restriction between the thorax and abdomen, and many of them possess stings. Many families have been recorded in amber. The Apocrita are divided into the Aculeata and the Parasitica.

Aculeata

The aculeates are generally large, have four straight veins originating from the base of the wing and several cells in the centre of the wing. The superfamily Apoidea comprise the bees. Bees are easy to identify in that they are fat and usually hairy. The front of the head is flat with two vertically elongate eyes and antennae that bend at 90 degrees in the middle. Bees are very rare in Baltic amber, but there is a species that is common in Dominican amber – *Proplebeia dominicana* (fig 89). This is a small stingless bee (4mm long) belonging to the family Apidae that collected tree resin to make its nest and often got trapped while doing this. Sometimes you can see the pollen sacs on its legs, which are full of resin. Bees are important pollinators of flowering plants. The first fossil orchid bee (subfamily Euglossinae) was recently discovered in Colombian copal and is a striking metallic green colour (fig 147). A closely related family to the bees is the Sphecidae. Sphecids are commonly known as digger wasps and like bees they too are pollinators of flowers. These wasps are more slender than bees and not hairy (fig 148). The forewings usually have two small, almost square cells towards the tip of the wing. Sphecids are very rare in amber.

The Vespidae are the social wasps. Today we recognize them as the large yellow-and-black striped insects that like jam and can sting! Their wings can fold longitudinally in half when at rest, which prevents them from damaging their wings while in the nest (fig 149). They are extremely rare in amber.

The Formicidae comprise the ants, which are the most abundant insects in Dominican amber. They are also

147 Orchid bee (Aculeata: Apidae: Euglossinae) in Colombian copal. (Length 10mm)

148 Digger wasp (Aculeata: Sphecidae) in Baltic amber. (Length 3.8mm)

149 Social wasp (Aculeata: Vespidae) in Dominican amber. (Length 13mm)

WASPS, BEES AND ANTS

150 Ant (Aculeata: Formicidae) in Baltic amber. (Length 4.1mm)

common in Baltic amber. This is not surprising as ants frequently run up and down tree trunks. Ants differ from other Hymenoptera in that they have two constrictions between the thorax and abdomen. They have squarish heads with small round eyes, large jaws and antennae that bend at 90 degrees in the middle. They, like some bees, are social insects with different castes. The workers are easy to identify in that they do not have wings (fig 150). Winged ants (kings and queens) have a distinctive venation. Most have a cross-shape formed by two veins intersecting towards the tip of the wing and many also have a square cell in the centre of the wing, orientated in a diamond shape (fig 151).

Parasitica

The Parasitica are generally much smaller than aculeates. Most are only 2mm or less in length and have a single unbranched vein along the anterior edge of the wing. Many different families have been recorded in amber but they require a powerful microscope to examine them and they are very difficult to tell apart by a non-expert.

The largest and most distinctive of the parasitic wasps belong to the superfamily Ichneumonoidea. These have slender bodies, long antennae and often have a long pointed ovipositor projecting from the tip of the abdomen. Those with long ovipositors parasitise larvae living in wood. The wing venation is less reduced than in other parasitic wasps with three straight veins originating from the base of the forewing. The superfamily consists of

151 Flying ant (Aculeata: Formicidae) in Baltic amber. (Length 6.5mm)

152 Ichneumon wasp (Parasitica: Ichneumonidae) in Baltic amber. (Length 6mm)

two families. The Ichneumonidae (ichneumon wasps) generally have two cells below the pterostigma (fig 152). One of the cells is large and curved and may have an additional tiny cell at its tip. The Braconidae (braconid wasps) by contrast have three cells below the pterostigma (fig 153). Ichneumonoids are rare in amber.

The Mymaridae and the Mymarommatidae are commonly known as fairy flies because they have a rim of long hairs around the edge of the wing. Mymarids have four elongate wings whereas mymarommatids have only two rounded wings. Both are parasites of the eggs of other insects. The mymarids are the smallest known insects; the smallest living today is 0.2mm long and the smallest recorded in amber (Dominican) is 0.25mm long (fig 154). Mymarommatids are very primitive and specimens recorded from Baltic amber are indistinguishable from specimens of

a living species. Fairy flies are very rare in amber, but this is probably because they are small and easily overlooked. Parasitic wasps as a whole are fairly rare in Baltic amber but fairly common in Dominican amber.

The mymarids belong to the superfamily Chalcidoidea. Some of the other members of this superfamily differ from other apocritans in that they do not have the characteristic constriction between thorax and abdomen (fig 155).

154 Fairy fly (Parasitica: Mymaridae) in Dominican amber. (Length 0.25mm)

153 Braconid wasp (Parasitica: Braconidae) in Baltic amber. (Length excluding ovipositor 3.4mm)

155 Chalcid wasp (Parasitica: Chalcidoidea) in Baltic amber. (Length 1mm)

Key to wasps, bees and ants

81 How many wings does it have?

0 Formicidae (ant) c/vc fig 150

2 (with a fringe of hairs) Mymarommatidae (fairy fly) VR

4 go to 82

82 Does it have a constriction between thorax and abdomen?

YES Apocrita go to 83

NO go to 91

KEY TO WASPS, BEES AND ANTS

83 Are its wings folded in half longitudinally?

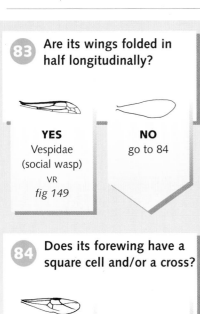

YES
Vespidae
(social wasp)
VR
fig 149

NO
go to 84

84 Does its forewing have a square cell and/or a cross?

YES
Formicidae
(ant)
C/VC
fig 151

NO
go to 85

85 Is its body short and fat?

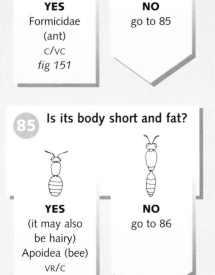

YES
(it may also be hairy)
Apoidea (bee)
VR/C
fig 89

NO
go to 86

86 Does it have four veins originating from the base of the forewing?

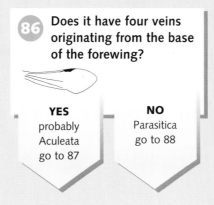

YES
probably
Aculeata
go to 87

NO
Parasitica
go to 88

87 Does it have two small cells near the tip of the wing?

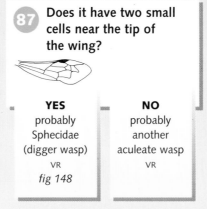

YES
probably
Sphecidae
(digger wasp)
VR
fig 148

NO
probably
another
aculeate wasp
VR

88 Does it have three veins projecting from the base of the forewing?

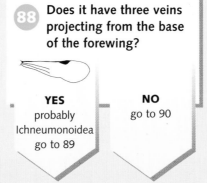

YES
probably
Ichneumonoidea
go to 89

NO
go to 90

89 Does it have a large curved cell in the centre of the wing?

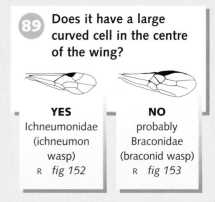

YES
Ichneumonidae
(ichneumon
wasp)
R *fig 152*

NO
probably
Braconidae
(braconid wasp)
R *fig 153*

90 Is it very small (about 1mm or less) and does it have wings surrounded by long hairs?

YES
probably
Mymaridae
(fairy fly)
VR *fig 154*

NO
another
parasitic wasp
R/C

91 Does the wing only have a single unbranched vein?

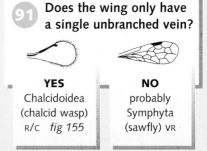

YES
Chalcidoidea
(chalcid wasp)
R/C *fig 155*

NO
probably
Symphyta
(sawfly) VR

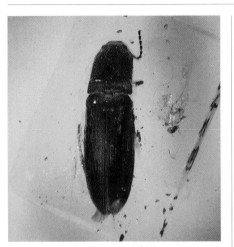

156 Click beetle (Coleoptera: Elateroidea) in Baltic amber. (Length 8mm)

157 Flat-footed beetle (Coleoptera: Platypodidae) in Dominican amber. (Length 3.2mm)

158 Rove beetle (Coleoptera: Staphylinidae) in Baltic amber. (Length 2.3mm)

COLEOPTERA – BEETLES

The order Coleoptera comprise the beetles. They are characterized by having a pair of highly toughened wing-cases (elytra) that meet down the middle of the body when at rest. This is the most diverse group of insects living today with about 400,000 described species and there are also many undescribed species. Beetles as a whole are fairly common in amber and more than 70 families have been recorded; however, each family is generally rare. In Baltic amber they often have a white coating, which unfortunately hides parts of the beetle and makes identification difficult. Furthermore, most families of beetles appear very similar to each other and are difficult to identify by a non-expert. Most beetles in amber are only a

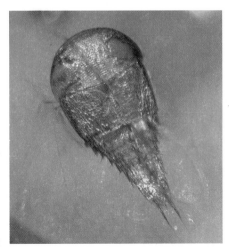

159 Rove beetle (Staphylinidae) in Dominican amber. This one would have lived inside termite nests. (Length 2mm)

couple to several millimetres in length. Larger beetles are extremely rare, presumably because they are strong and pull themselves out of sticky resin.

There are four suborders of Coleoptera, three of which occur in amber. The majority of beetles belong to the Polyphaga.

POLYPHAGA

The Polyphaga have diverse feeding habits. Given that there are so many similar-looking families recorded in amber, only four distinctive superfamilies and families are described here.

The most common beetles in Baltic amber belong to the superfamily Elateroidea, which includes the click beetles. They have long bodies with pointed and striated elytra (wing-cases) (fig 156). The corners of the thorax are pointed where they meet the elytra. Elateroids are wood-boring beetles that feed on rotting or diseased wood. They are very rare in Dominican amber.

The most common beetles in Dominican amber belong to the family Platypodidae. They are known as flat-footed beetles, but their feet are not noticeably flat! They have a long cylindrical body with an elongate thorax, widely spaced legs and short antennae with an enlarged terminal segment (fig 157). Platypodids are also wood-borers, which probably bored into the *Hymenaea* trees that produced the Dominican amber. They are extremely rare in Baltic amber.

BEETLES

The Staphylinidae are known commonly as rove beetles or devil's coach horses. They have elongate bodies with small elytra (fig 158). They look like earwigs (Dermaptera) except they do not have pincers. They have large jaws and are voracious predators that feed on other invertebrates. There are unusual specimens present in Dominican amber that are small, have a large semicircular head, short antennae and a short tapering body (fig 159). These beetles are specially adapted to living inside termite nests.

The superfamily Curculionoidea comprise the weevils. They are easy to identify in that they have a snout (fig 160). Their antennae originate on the snout and usually bend in the middle. Weevils feed on plants.

160 Weevil (Coleoptera: Curculionoidea) in Dominican amber. (Length 1.7mm)

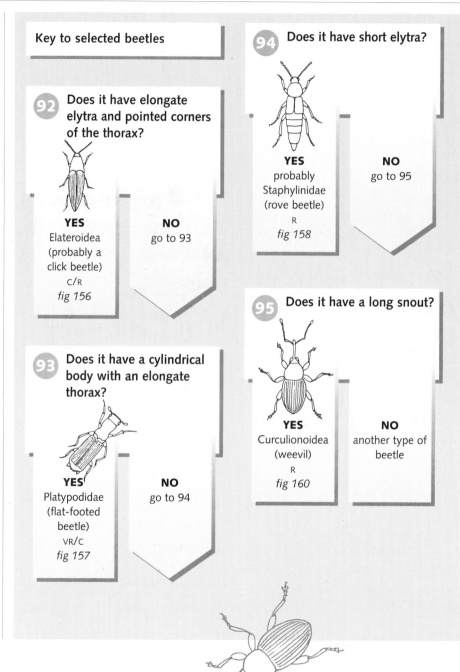

Key to selected beetles

92 **Does it have elongate elytra and pointed corners of the thorax?**

YES
Elateroidea (probably a click beetle)
C/R
fig 156

NO
go to 93

93 **Does it have a cylindrical body with an elongate thorax?**

YES
Platypodidae (flat-footed beetle)
VR/C
fig 157

NO
go to 94

94 **Does it have short elytra?**

YES
probably Staphylinidae (rove beetle)
R
fig 158

NO
go to 95

95 **Does it have a long snout?**

YES
Curculionoidea (weevil)
R
fig 160

NO
another type of beetle

161 *Large beetle (Coleoptera: Elateroidea?) in Burmese amber. (Length 19mm)*

GLOSSARY

Abdomen The hind-body of an animal, containing the organs for digestion and reproduction.

Amber Fossilized tree resin that is usually yellow or orange in colour and is commonly used for jewellery.

Amberization The process whereby tree resin fossilizes into amber due to hardening, the loss of volatiles and polymerization.

Amorphous Does not have a definite form or crystalline structure.

Ambroid A substance made by fusing small pieces of amber using heat and pressure. Also known as pressed or reconstituted amber.

Apterygotes Insects that do not have wings.

Arista A bristle that projects from the antennae of some flies.

Binomial nomenclature The scientific naming of organisms. Every organism is given a genus and species name in Latin.

Biogeography The study of the geographical distribution of organisms.

Carapace The toughened shell of an animal, such as the head-shield of Crustacea.

Caudal filaments The long, thin tails of some insects, such as mayflies.

Cerci A pair of sensory appendages that project from the tip of the abdomen of some insects, e.g. crickets.

Cephalothorax The fused head and thorax of some invertebrate animals, such as spiders.

Chelicerae A pair of curved, downward pointing fangs of some invertebrate animals, such as spiders.

Chrysalis See pupa.

Clavus A distinctive rounded or triangular area at the base of the forewings of some insects, such as cockroaches.

Colophony The solid residue from the distillation of amber to extract acid and oil.

Commensalism An association between two organisms in which one benefits and the other derives no benefit or harm.

Conchoidal fracture A type of breakage in which the surface resembles a shell.

Copal Hardened tree resin that is modern or sub-fossil. It is usually less than 2 million years old.

Costa A vein of an insect's wing. It forms the front (anterior) edge of the wing.

Cubitus One of the main branching veins of an insect's wing. It usually terminates on the hind (posterior) edge of the wing.

Diversity The variety of life.

DNA Deoxyribonucleic acid, the molecule that carries the genetic information essential for life.

Ecology The study of the interactions between animals, plants and their physical surroundings.

Ectognathous Insects with mouthparts that project externally from the head.

Endopterygotes Insects that undergo complete metamorphoses during their life cycle. The egg hatches into a larva, which pupates and metamorphoses into an adult.

Entognathous Insects with mouthparts that are enclosed within the head.

Entomologist Someone who studies insects.

Entomology The study of insects.

Epiphytes Plants that grow attached to other plants.

Evolution The gradual development of organisms from other more primitive organisms.

Exopterygotes Insects that undergo incomplete metamorphosis during their life cycle. The egg hatches into a nymph, which grows into an adult.

Extant Living today.

Extinct Died out.

Fluoresce To glow when subjected to invisible light of a short wavelength, such as ultraviolet.

Fossil The remains of dead organisms, usually preserved in rock.

Genus A taxonomic unit consisting of a group of similar species.

Halteres Small knob-like organs of flies, used as balancers during flight and which are highly reduced hindwings.

Hydrocarbon An organic molecule containing hydrogen and carbon.

Infra-red spectroscopy A method of analysing amber whereby infrared light is passed through a compressed pellet of powdered amber and potassium bromide. The frequencies that are absorbed are plotted on a graph.

In situ In its original place.

Inclusion Something that is trapped in amber, such as insects, plant remains and bubbles.

Larva A juvenile crawling stage of an insect that undergoes complete metamorphosis. Also known as a caterpillar, grub or maggot.

Mass spectrometry A method of analysing amber whereby a small amount of amber is vaporized. The resultant molecules are then charged and passed through a magnetic field, which changes their path according to their mass. The spectrometer then measures these changes and plots them on a graph.

Median One of the main branching veins of an insect's wing, which terminates at the wing tip and/or the hind (posterior) edge of the wing.

Metamorphosis The transformation of a juvenile into an adult. For insects with incomplete metamorphosis the

GLOSSARY

juvenile (nymph) grows wings to become the adult. For insects with complete metamorphosis the juvenile (larva) pupates, then the tissues break down and re-form as the adult.

Morphology The form (appearance) of an organism.

Moulting The shedding of skin to enable an organism to grow.

Mutualism An association between two organisms in which both derive benefit.

Nymph A juvenile stage of an insect that undergoes incomplete metamorphosis. The nymph generally looks like the adult but is smaller and lacks fully developed wings.

Organic Something that contains carbon and is produced by living things.

Palaeoecology The study of the relations of animals and plants to one another and their physical surroundings in the distant past.

Palaeontology The study of fossils.

Parasite An organism living in or on another and benefiting at the expense of the other.

Pedipalps A pair of appendages attached to the head of some animals, which have either a sensory function, such as in spiders, or are used to gather food, such as the pincers of scorpions.

Phoresy An association between two organisms where one is carried by the other, but is not a parasite.

Plate tectonics The study of the earth's surface, which is subdivided into continental and oceanic plates that move with time.

Polymer A large organic molecule composed of many smaller molecules.

Polymeric Something that is composed of polymers.

Polymerisation The process whereby molecules join together to form larger molecules (polymers).

Population The inhabitants of a particular area.

Proboscis The elongated mouthparts of some insects, which are formed into a tube and used for sucking up liquids, such as nectar by butterflies, blood by mosquitoes or sap by aphids.

Pronotum The toughened head-shield of some insects, such as cockroaches.

Pterostigma A dark spot found near the tip of the wings of some insects.

Pupa A toughened case produced by insects that undergo complete metamorphosis, in which the larval tissues break down and re-form to produce an adult insect. Also known as a chrysalis or cocoon.

Pyrite A mineral made up of iron and sulphur. Also known as Fool's Gold because of its colour.

Radial sector The posterior branch of the radius vein.

Radius One of the main veins of an insect's wing, which usually terminates at the wing tip. It usually runs parallel to the subcosta and branches posteriorly into the radial sector.

Refractive index The ratio of the speed of light when it passes through a transparent material compared to its speed when it passes through a vacuum.

Resin A sticky organic substance produced by trees for defence against disease and insect attack.

Rostrum The elongate beak-like mouthparts of some insects, such as scorpionflies.

Sclerotized Toughened.

Sexual dimorphism The differences in form between the males and females of a species.

Solvent A substance that is able to dissolve other substances.

Speciation The formation of a new species from another by evolution.

Species A group of individuals with similar characteristics, which are able to reproduce with one another.

Specific gravity The weight of a substance measured as a ratio of its density to either water (for solids and liquids) or air (for gases). Also known as relative density.

Stellate hairs Tiny star-like clusters of hairs that are commonly found in Baltic amber and probably came from the male flowers of oak trees.

Subcosta A vein of an insect wing, which is usually simple and runs parallel to the front (anterior) edge of the wing.

Sun spangles Circular silvery cracks in amber, particularly common in modern amber jewellery. They are artificially made by heating the amber.

Taxa The units of classification.

Taxonomy The science of naming and classifying organisms.

Tegmina Toughened forewings, which are used for protection by some insects, such as cockroaches.

Thorax The middle section of the body of some animals, such as insects.

Type specimen The original specimen that was used as the basis for the description and naming of a species.

Venation The pattern of veins in the wings of insects.

Viscous A viscous substance is sticky and unable to flow freely.

Volatiles Liquids that evaporate easily.

INDEX